Workshop on Medical Speech Translation 2006

Held at HLT/NAACL 2006

New York, New York, USA
9 June 2006

ISBN: 978-1-61738-496-7

Printed from e-media with permission by:

Curran Associates, Inc.
57 Morehouse Lane
Red Hook, NY 12571
www.proceedings.com

Some format issues inherent in the e-media version may also appear in this print version.

Printed by Curran Associates, Inc. (2010)

Association for Computational Linguistics
209 N. Eighth Street
Stroudsburg, Pennsylvania 18360

Phone: 1-570-476-8006
Fax: 1-570-476-0860

acl@aclweb.org

Introduction

Medical applications have emerged as one of the most popular domains for speech translation, and several functional systems now exist. Despite this, there is so far no established consensus on any of the central questions, including the following:

- Does medical speech translation pose special problems, and if so, what are they?
- What do the users (both doctors and patients) actually want? What constitutes acceptable performance, given that medicine is a safety-critical area?
- What are the alternatives to speech translation for non-L1 speakers in healthcare situations?
- What are the most important tasks, sub-domains and language pairs?
- What architectures are most suitable for medical speech translation applications? (Fixed-phrase, ad hoc phrasal rules, rule-based, statistical...)
- What evaluation/data collection methodologies are appropriate to medical speech translation?
- What requirements are there on hardware platforms? What options currently exist?
- How close are we to having applications that can be used in the field?

In this one day workshop, our aim has been to get together as many as possible of the key players in this field, so that we can exchange information and clarify the above and other issues. We expect the workshop to be of interest to people working in all three component communities - speech technology, machine translation, and medicine.

The main body of the workshop consists of two parts: oral presentation of papers, followed by a demo session. We will end with a panel discussion, which will include representatives of both the system developer and medical user communities.

Organizers:

Pierrette Bouillon University of Geneva (Switzerland)
Farzad Ehsani Sehda, Inc. (US)
Robert Frederking Carnegie Mellon University (US)
Manny Rayner University of Geneva (Switzerland); ICSI/NASA Ames Research Center (US)

Program Committee:

Hervé Blanchon CLIPS-GETA, Grenoble (France)
Pierrette Bouillon University of Geneva (Switzerland)
Mike Dillinger Spokentranslation.com (US)
Farzad Ehsani Sehda, Inc. (US)
Glenn Flores Medical College of Wisconsin (US)
Robert Frederking Carnegie Mellon University (US)
John Fry Stanford University (US)
John Hutchins University of East Anglia (UK)
Hitoshi Isahara NICT (Japan)
Lori Levin Carnegie Mellon University (US)
Shri Narayanan USC Viterbi School of Engineering (US)
Manny Rayner University of Geneva (Switzerland); ICSI/NASA Ames Research Center (US)
Harold Somers University of Manchester (UK)
Tanja Schultz Carnegie Mellon University (US)
Vol Van Dalsem III El Camino Hospital (US)
Bowen Zhou IBM, T.J. Watson Research Center (US)

Table of Contents

Conference Program

9:15–9:30 Opening Remarks

Session 1: Full Papers 1

9:30–10:00 *Usability Issues in an Interactive Speech-to-Speech Translation System for Healthcare*
Mark Seligman and Mike Dillinger

10:00–10:30 *Evaluating Task Performance for a Unidirectional Controlled Language Medical Speech Translation System*
Nikos Chatzichrisafis, Pierrette Bouillon, Manny Rayner, Marianne Santaholma, Marianne Starlander and Beth Ann Hockey

10:30–11:00 Break

Session 2: Full Papers 2

11:00–11:30 *Speech to Speech Translation for Medical Triage in Korean*
Farzad Ehsani, Jim Kimzey, Demetrios Master, Karen Lesea and Hunil Park

11:30–12:00 *Automated Interpretation of Clinical Encounters with Cultural Cues and Electronic Health Record Generation*
Daniel T. Heinze, Alexander Turchin and V. Jagannathan

12:00–12:30 *Language Engineering and the Pathway to Healthcare: A User-Oriented View*
Harold Somers

12:30–14:00 Lunch

Session 3: Formal Demo Presentations

14:00–14:15 *Converser (TM): Highly Interactive Speech-to-Speech Translation for Healthcare*
Mike Dillinger and Mark Seligman

14:15–14:30 *MedSLT: A Limited-Domain Unidirectional Grammar-Based Medical Speech Translator*
Manny Rayner, Pierrette Bouillon, Nikos Chatzichrisafis, Marianne Santaholma, Marianne Starlander, Beth Ann Hockey, Yukie Nakao, Hitoshi Isahara and Kyoko Kanzaki

14:30–14:45 *S-MINDS 2-Way Speech-to-Speech Translation System*
Farzad Ehsani, Jim Kimzey, Demetrios Master, Karen Sudre, David Domingo and Hunil Park

14:45–15:00 *Accultran: Automated Interpretation of Clinical Encounters with Cultural Cues and Electronic Health Record Generation*
Daniel T. Heinze, Alexander Turchin and V. Jagannathan

15:00–15:15 *A Multi-Lingual Decision Support Prototype for the Medical Domain*
David Dinh, Dennis Chan and Jack Chen

15:15–15:30 *IBM MASTOR SYSTEM: Multilingual Automatic Speech-to-Speech Translator*
Yuqing Gao, Bowen Zhou, Ruhi Sarikaya, Mohamed Afify, Hong-Kwang Kuo, Wei-zhong Zhu, Yonggang Deng, Charles Prosser, Wei Zhang and Laurent Besacier

15:30–16:00 Break

Session 4: Informal Demo Presentations and Panel

16:00–17:00 Demos
17:00–17:30 Panel

Usability Issues in an Interactive Speech-to-Speech Translation System for Healthcare

Mark Seligman
Spoken Translation, Inc.
Berkeley, CA, USA 94705
mark.seligman@spokentranslation.com

Mike Dillinger
Spoken Translation, Inc.
Berkeley, CA, USA 94705
mike.dillinger@spokentranslation.com

Abstract

We describe a highly interactive system for bidirectional, broad-coverage spoken language communication in the healthcare area. The paper briefly reviews the system's interactive foundations, and then goes on to discuss in greater depth issues of practical usability. We present our Translation Shortcuts facility, which minimizes the need for interactive verification of sentences after they have been vetted once, considerably speeds throughput while maintaining accuracy, and allows use by minimally literate patients for whom any mode of text entry might be difficult. We also discuss facilities for multimodal input, in which handwriting, touch screen, and keyboard interfaces are offered as alternatives to speech input when appropriate. In order to deal with issues related to sheer physical awkwardness, we briefly mention facilities for hands-free or eyes-free operation of the system. Finally, we point toward several directions for future improvement of the system.

1 Introduction

Increasing globalization and immigration have led to growing demands on US institutions for healthcare and government services in languages other than English. These institutions are already overwhelmed: the State of Minnesota, for example, had no Somali-speaking physicians for some 12,000 Somali refugees and only six Hmong-speaking physicians to serve 50,000 Hmong residents (Minnesota Interpreter Standards Advisory Committee, 1998). San Francisco General Hospital, to cite another example, receives approximately 3,500 requests for interpretation per month, or 42,000 per year for 35 different languages. Moreover, requests for medical interpretation services are distributed among all the wards and clinics, adding a logistical challenge to the problem of a high and growing demand for interpretation services (Paras, et al., 2002). Similar situations are found throughout the United States.

It is natural to hope that automatic real-time translation in general, and spoken language translation (SLT) in particular, can help to meet this communicative need. From the viewpoint of research and development, the high demand in healthcare makes this area especially attractive for fielding early SLT systems and seeking early adopters.

With this goal in view, several speech translation systems have aimed at the healthcare area. (See www.sehda.com, DARPA's CAST program, www.phraselator.com, etc.) However, these efforts have encountered several issues or limitations.

First, they have been confined to narrow domains. In general, SLT applications have been able to achieve acceptable accuracy only by staying within restricted topics, in which fixed phrases could be used (e.g., www.phraselator.com), or in which grammars for automatic speech recognition (ASR) and machine translation (MT) could be optimized. For example, MedSLT (Bouillon et al, 2005) is limited to some 600 specific words per sub-domain. IBM's MASTOR system, with 30,000 words in each translation direction, has much broader coverage, but remains comparable in lexicon size to commercial MT systems of the early 1980s.

Granted, restriction to narrow domains may often be appropriate, given the large effort involved in compiling extensive lexical resources and the time required for deployment. A tightly focused approach permits relatively quick development of new systems and provides a degree of flexibility to experiment with different architectures and different languages.

Our emphasis, however, is on breaking out of narrow domains. We seek to maximize versatility by providing exceptional capacity to move from topic to topic while maintaining adequate accuracy.

To provide a firm foundation for such versatility, we "give our systems a liberal arts education" by incorporating very broad-coverage ASR and MT technology. Our MT lexicons, for example, contain roughly 300,000 words in each direction.

But of course, as coverage increases, perplexity and the ASR and MT errors due to it increase in proportion, especially in the absence of tight integration between these components. To compensate, we provide a set of facilities that enable users from both sides of the language barrier to interactively monitor and correct these errors. Putting users in the speech translation loop in this way does in fact permit conversations to range widely (Seligman, 2000). We believe that this highly interactive approach will prove applicable to the healthcare area.

We have described these interactive techniques in (Dillinger and Seligman, 2004; Zong and Seligman, forthcoming). We will review them only briefly here, in Section 2.

A second limitation of current speech translation systems for healthcare is that bilingual (bidirectional) communication has been difficult to enable. While speech-to-speech translation has sometimes proven practical from the English side, translation from the non-English side has been more difficult to achieve. Partly, this limitation arises from human factors issues: while naïve observers might expect spoken input to be effortless for anyone who can talk, the reality is that users must learn to use most speech interfaces, and that this learning process can be difficult for users who are less literate or less computer literate. Further, many healthcare venues make speech input difficult: they may be noisy, microphones may be awkward to situate or to pass from speaker to speaker, and so on.

Our group's approach to training- or venue-related difficulties for speech input is to provide an array of alternative input modes. In addition to providing input through dictated speech, users of our system can freely alternate among three other input modes, using handwriting, a touch screen, and standard bilingual keyboards.

In this paper, we will focus on practical usability issues in the design of user interfaces for highly interactive approaches to SLT in healthcare applications. With respect to interactivity per se, we will discuss the following specific issues:

- In a highly interactive speech translation system, monitoring and correction of ASR and MT are vital for accuracy and confidence, but can be time consuming – in a field where time is always at a premium.
- Interactivity demands a minimum degree of computer and print literacy, which some patients may lack.

To address these issues, we have developed a facility called *Translation Shortcuts*™, to be explained throughout Section 3.

Section 4 will describe our approach to multimodal input. As background, however, Section 2 will quickly review our approach to highly interactive – and thus uniquely broad-coverage – spoken language translation. Before concluding, we will in Section 5 point out planned future developments.

2 Highly Interactive, Broad-coverage SLT

We now briefly summarize our group's approach to highly interactive, broad-coverage SLT.

The twin goals of accuracy and broad-coverage have generally been in opposition: speech translation systems have gained tolerable accuracy only by sharply restricting both the range of topics that can be discussed and the sets of vocabulary and structures that can be used to discuss them. The essential problem is that both speech recognition and translation technologies are still quite error-prone. While the error rates may be tolerable when each technology is used separately, the errors combine and even compound when they are used together. The resulting translation output is generally below the threshold of usability – unless restriction to a very narrow domain supplies sufficient constraints to significantly lower the error rates of both components.

As explained, our group's approach has been to concentrate on interactive monitoring and correction of both technologies.

First, users can monitor and correct the speaker-dependent speech recognition system to ensure that the text that will be passed to the machine translation component is completely correct. Voice commands (e.g. **Scratch That** or **Correct <incorrect text>**) can be used to repair speech recognition errors. Thus, users of our SLT enrich the interface between ASR and MT.

Next, during the MT stage, users can monitor, and if necessary correct, one especially important aspect of the translation – lexical disambiguation.

Our system's approach to lexical disambiguation is twofold: first, we supply a *Back-Translation*, or re-translation of the translation. Using this paraphrase of the initial input, even a monolingual user can make an initial judgment concerning the quality of the preliminary machine translation output. (Other systems, e.g. IBM's MASTOR, have also employed re-translation. Our implementations, however, exploit proprietary technologies to ensure that the lexical senses used during back translation accurately reflect those used in forward translation.)

In addition, if uncertainty remains about the correctness of a given word sense, we supply a proprietary set of Meaning Cues™ – synonyms, definitions, etc. – which have been drawn from various resources, collated in a database (called SELECT™), and aligned with the respective lexica of the relevant MT systems. With these cues as guides, the user can monitor the current, proposed meaning and select (when necessary) a different, preferred meaning from among those available. Automatic updates of translation and back translation then follow. Future versions of the system will allow personal word-sense preferences thus specified in the current session to be stored and reused in future sessions, thus enabling a gradual tuning of word-sense preferences to individual needs. Facilities will also be provided for sharing such preferences across a working group.

Given such interactive correction of both ASR and MT, wide-ranging, and even jocular, exchanges become possible (Seligman, 2000).

As we have said, such interactivity within a speech translation system can enable increased accuracy and confidence, even for wide-ranging conversations.

Accuracy of translation is, in many healthcare settings, critical to patient safety. When a doctor is taking a patient's history or instructing the patient in a course of treatment, even small errors can have clinically relevant effects. Even so, at present, healthcare workers often examine patients and instruct them in a course of treatment through gestures and sheer good will, with no translation at all, or use untrained human interpreters (friends, family, volunteers, or staff) in an error-prone attempt to solve the immediate problem (Flores, et al., 2003). As a result, low-English proficiency patients are often less healthy and receive less effective treatment than English speakers (Paras, et al., 2002). We hope to demonstrate that highly interactive real-time translation systems in general, and speech translation systems in particular, can help to bridge the language gap in healthcare when human interpreters are not available.

Accuracy in an automatic real-time translation system is necessary, but not sufficient. If healthcare workers have no means to independently assess the reliability of the translations obtained, practical use of the system will remain limited. Highly interactive speech translation systems can foster the confidence on both sides of the conversation, which is necessary to bring such systems into wide use. In fact, in this respect at least, they may sometimes prove superior to human interpreters, who normally do not provide clients with the means for judging translation accuracy.

The value of enabling breadth of coverage, as well as accuracy and confidence, should also be clear: for many purposes, the system must be able to translate a wide range of topics *outside of* the immediate healthcare domain – for example, when a patient tries to describe what was going on when an accident occurred. The ability to ask about interests, family matters, and other life concerns is vital for establishing rapport, managing expectations and emotions, etc.

3 Translation Shortcuts

Having summarized our approach to highly interactive speech translation, we now turn to examination of practical interface issues for this class of SLT system. This section concentrates on Translation Shortcuts™.

Shortcuts are designed to provide two main advantages:

First, re-verification of a given utterance is unnecessary. That is, once the translation of an utterance has been verified interactively, it can be saved for later reuse, simply by activating a **Save as Shortcut** button on the translation verification screen. The button gives access to a dialogue in which a convenient *Shortcut Category* for the Shortcut can be selected or created. At reuse time, no further verification will be required. (In addition to such dynamically created *Personal* Shortcuts, any number of prepackaged *Shared* Shortcuts can be included in the system.)

Second, access to stored Shortcuts is very quick, with little or no need for text entry. Several facilities contribute to meeting this design criterion.

- A *Shortcut Search* facility can retrieve a set of relevant Shortcuts given only keywords or the first few characters or words of a string. The desired Shortcut can then be executed with a single gesture (mouse click or stylus tap) or voice command.

  NOTE: If no Shortcut is found, the system automatically allows users access to the full power of broad-coverage, interactive speech translation. Thus, a seamless transition is provided between the Shortcuts facility and full, broad-coverage translation.

- A *Translation Shortcuts Browser* is provided, so that users can find needed Shortcuts by traversing a tree of Shortcut categories. Using this interface, users can execute Shortcuts even if their ability to input text is quite limited, e.g. by tapping or clicking alone.

Figure 1 shows the Shortcut Search and Shortcuts Browser facilities in use. Points to notice:

- On the left, the Translation Shortcuts Panel has slid into view and been pinned open. It contains the Translation Shortcuts Browser, split into two main areas, Shortcuts Categories (above) and Shortcuts List (below).
- The Categories section of the Panel shows current selection of the **Conversation** category, containing everyday expressions, and its **Staff** subcategory, containing expressions most likely to be used by healthcare staff members. There is also a **Patients** subcategory, used for patient responses. Categories for **Administrative topics** and **Patient's Current Condition** are also visible; and new ones can be freely created.
- Below the Categories section is the Shortcuts List section, containing a scrollable list of alphabetized Shortcuts. (Various other sorting criteria will be available in the future, e.g. sorting by frequency of use, recency, etc.)
- Double clicking on any visible Shortcut in the List will execute it. Clicking once will select and highlight a Shortcut. Typing **Enter** will execute the currently highlighted Shortcut (here "Good morning"), if any.
- It is possible to automatically relate options for a patient's response to the previous staff member's utterance, e.g. by automatically going to the sibling **Patient** subcategory if the prompt was given from the **Staff** subcategory.

Because the Shortcuts Browser can be used without text entry, simply by pointing and clicking, it enables responses by minimally literate users. In the future, we plan to enable use even by completely illiterate users, through two devices: we will enable automatic pronunciation of Shortcuts and categories in the Shortcuts Browser via text-to-speech, so that these elements can in effect be read aloud to illiterate users; and we will augment Shared Shortcuts with pictorial symbols, as clues to their meaning.

A final point concerning the Shortcuts Browser: it can be operated entirely by voice commands, although this mode is more likely to be useful to staff members than to patients.

We turn our attention now to the Input Window, which does double duty for Shortcut Search and arbitrary text entry for full translation. We will consider the search facility first, as shown in Figure 2.

- Shortcuts Search begins automatically as soon as text is entered by any means – voice, handwriting, touch screen, or standard keyboard – into the Input Window.
- The **Shortcuts Drop-down Menu** appears just below the Input Window, as soon as there are results to be shown. The user has entered "Good" and a space, so the search program has received its first input word. The drop-down menu shows the results of a keyword-based search.
- Here, the results are sorted alphabetically. Various other sorting possibilities may be useful: by frequency of use, proportion of matched words, etc.

- The highest priority Shortcut according to the specified sorting procedure can be highlighted for instant execution.
- Other shortcuts will be highlighted differently, and both kinds of highlighting are synchronized with that of the Shortcuts list in the Shortcuts Panel.
- Arrow keys or voice commands can be used to navigate the drop-down list.
- If the user goes on to enter the exact text of any Shortcut, e.g. "Good morning," a message will show that this is in fact a Shortcut, so that verification will not be necessary. However, final text not matching a Shortcut, e.g. "Good job," will be passed to the routines for full translation with verification.

4 Multimodal input

As mentioned, an unavoidable issue for speech translation systems in healthcare settings is that speech input is not appropriate for every situation.

Current speech-recognition systems are unfamiliar for many users. Our system attempts to overcome this training issue to some extent by incorporating standard commercial-grade dictation systems for broad-coverage and ergonomic speech recognition. These products already have established user bases in the healthcare community. Even so, some training may be required: optional generic Guest profiles are supplied by our system for male and female voices in both languages; but optional voice enrollment, requiring five minutes or so, is helpful to achieve best results. Such training time is practical for healthcare staff, but will be realistic for patients only when they are repeat visitors, hospital-stay patients, etc.

As mentioned, other practical usability issues for the use of speech input in healthcare settings include problems of ambient noise (e.g. in emergency rooms or ambulances) and problems of microphone and computer arrangement (e.g. to accommodate not only desktops but counters or service windows which may form a barrier between staff and patient).

To deal with these and other usability issues, we have found it necessary to provide a range of input modes: in addition to dictated speech, we enable handwritten input, the use of touch screen keyboards for text input, and the use of standard keyboards. All of these input modes must be completely bilingual, and language switching must be arranged automatically when there is a change of active participant. Further, it must be possible to change input modes seamlessly within a given utterance: for example, users must be able to dictate the input if they wish, but then be able to make corrections using handwriting or one of the remaining two modes. Figure 3 shows such seamless bilingual operation: the user has dictated the sentence "Tengo náuseas" in Spanish, but there was a speech-recognition error, which is being corrected by handwriting.

Of course, even this flexible range of input options does not solve all problems. As mentioned, illiterate patients pose special problems. Again, naïve users tend to suppose that speech is the ideal input mode for illiterates. Unfortunately, however, the careful and relatively concise style of speech that is required for automatic recognition is often difficult to elicit, so that recognition accuracy remains low; and the ability to read and correct the results is obviously absent. Just as obviously, the remaining three text input modes will be equally ineffectual for illiterates.

As explained, our current approach to low literacy is to supply Translation Shortcuts for the minimally literate, and – in the future – to augment Shortcuts with text-to-speech and iconic pictures.

Staff members will usually be at least minimally literate, but they present their own usability issues.

Their typing skills may be low or absent. Handling the computer and/or microphone may be awkward in many situations, e.g. when examining a patient or taking notes. (Speech translation systems are expected to function in a wide range of physical settings: in admissions or financial aid offices, at massage tables for physical therapy with patients lying face down, in personal living rooms for home therapy or interviews, and in many other locations.)

To help deal with the awkwardness issues, our system provides voice commands, which enable hands-free operation. Both full interactive translation and the Translation Shortcut facility (using either the Browser or Search elements) can be run hands-free. To a limited degree, the system can be used *eyes*-free as well: text-to-speech can be used to pronounce the back-translation so that preliminary judgments of translation quality can be made without looking at the computer screen.

5 Future developments

We have already mentioned plans to augment the Translation Shortcuts facility with text-to-speech and iconic pictures, thus moving closer to a system suitable for communication with completely illiterate or incapacitated patients.

Additional future directions follow.

- **Server-based architectures:** We plan to move toward completely or partially server-based arrangements, in which only a very thin client software application – for example, a web interface – will run on the client device. Such architectures will permit delivery of our system on smart phones in the Blackberry or Treo class. Delivery on handhelds will considerably diminish the issues of physical awkwardness discussed above, and anytime/anywhere/any-device access to the system will considerably enlarge its range of uses.
- **Pooling Translation Shortcuts:** As explained above, the current system now supports both Personal (do-it-yourself) and Shared (prepackaged) Translation Shortcuts. As yet, however, there are no facilities to facilitate pooling of Personal Shortcuts among users, e.g. those in a working group. In the future, we will add facilities for exporting and importing shortcuts.
- **Translation memory:** Translation Shortcuts can be seen as a variant of Translation Memory, a facility that remembers past successful translations so as to circumvent error-prone reprocessing. However, at present, we save Shortcuts only when explicitly ordered. If all other successful translations were saved, there would soon be far too many to navigate effectively in the Translation Shortcuts Browser. In the future, however, we could in fact record these translations in the background, so that there would be no need to re-verify new input that matched against them. Messages would advise the user that verification was being bypassed in case of a match.
- **Additional languages:** The full SLT system described here is presently operational only for bidirectional translation between English and Spanish. We expect to expand the system to Mandarin Chinese next. Limited working prototypes now exist for Japanese and German, though we expect these languages to be most useful in application fields other than healthcare.
- **Testing**: Systematic usability testing of the full system is under way. We look forward to presenting the results at a future workshop.

6 Conclusion

We have described a highly interactive system for bidirectional, broad-coverage spoken language communication in the healthcare area. The paper has briefly reviewed the system's interactive foundations, and then gone on to discuss in greater depth issues of practical usability.

We have presented our Translation Shortcuts facility, which minimizes the need for interactive verification of sentences after they have been vetted once, considerably speeds throughput while maintaining accuracy, and allows use by minimally literate patients for whom any mode of text entry might be difficult.

We have also discussed facilities for multimodal input, in which handwriting, touch screen, and keyboard interfaces are offered as alternatives to speech input when appropriate. In order to deal with issues related to sheer physical awkwardness, we have briefly mentioned facilities for hands-free or eyes-free operation of the system.

Finally, we have pointed toward several directions for future improvement of the system.

References

Pierrette Bouillon, Manny Rayner, et al. 2005. A Generic Multi-Lingual Open Source Platform for Limited-Domain Medical Speech Translation. Presented at *EAMT 2005*, Budapest, Hungary.

Mike Dillinger and Mark Seligman. 2004. A highly interactive speech-to-speech translation system. *Proceedings of the VI Conference of the Association of Machine Translation in the Americas*. E. Stroudsburg, PA: American Association for Machine Translation.

Glenn Flores, M. Laws, S. Mays, et al. 2003. Errors in medical interpretation and their potential clinical consequences in pediatric encounters. *Pediatrics*, 111: 6-14.

Minnesota Interpreter Standards Advisory Committee. 1998. *Bridging the Language Gap: How to meet the need for interpreters in Minnesota*. Available at: http://www.cce.umn.edu/creditcourses/pti/downloads.html.

Melinda Paras, O. Leyva, T. Berthold, and R. Otake. 2002. *Videoconferencing Medical Interpretation: The results of clinical tri*als. Oakland, CA: Heath Access Foundation.

PHRASELATOR (2006). http://www.phraselator.com. As of April 3, 2006.

S-MINDS (2006). http://www.sehda.com/solutions.htm. As of April 3, 2006.

Mark Seligman. 2000. Nine Issues in Speech Translation. *Machine Translation,* 15, 149-185.

Chengqing Zong and Mark Seligman. Forthcoming. Toward Practical Spoken Language Translation. *Machine Translation.*

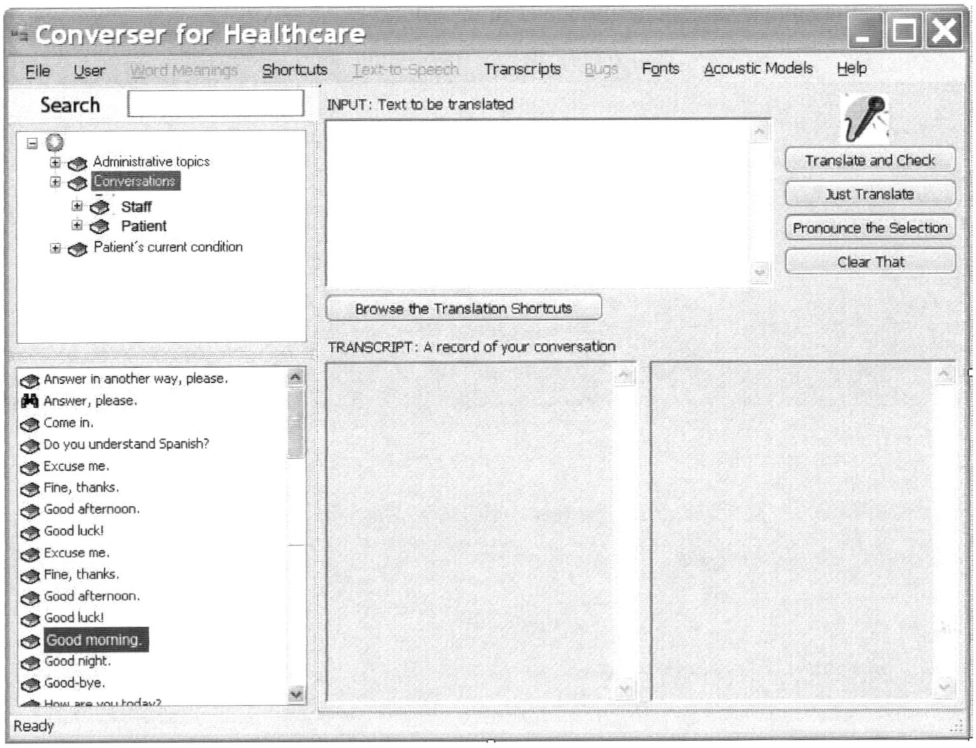

Figure 1: The Input Screen, showing the Translation Shortcuts Browser and Search facilities.

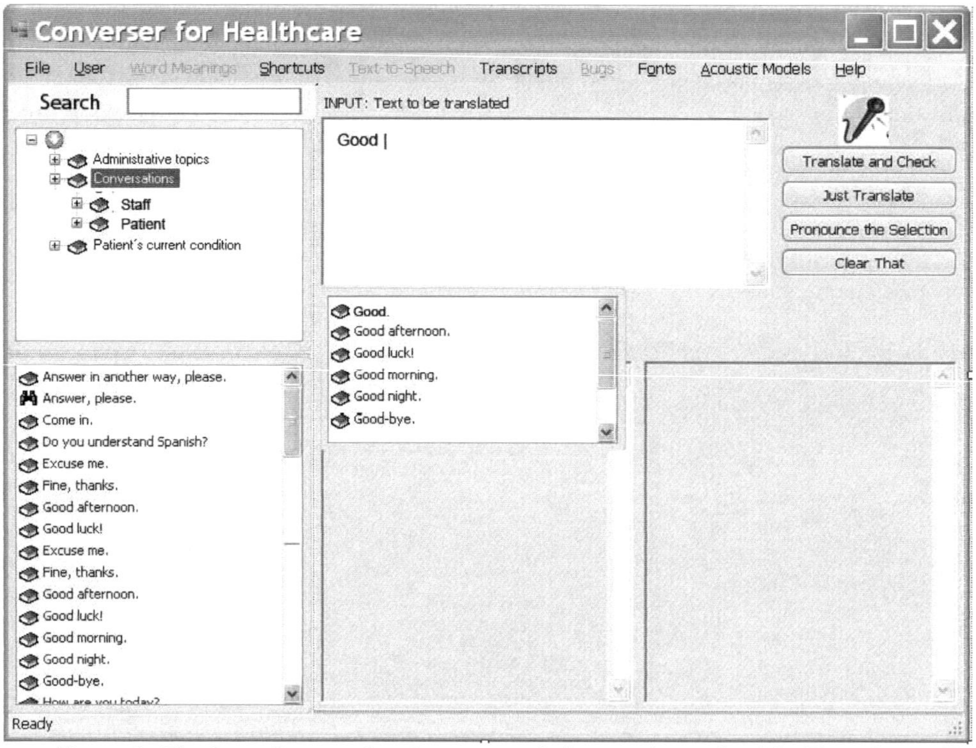

Figure 2: The Input Screen, showing automatic keyword search of the Translation Shortcuts.

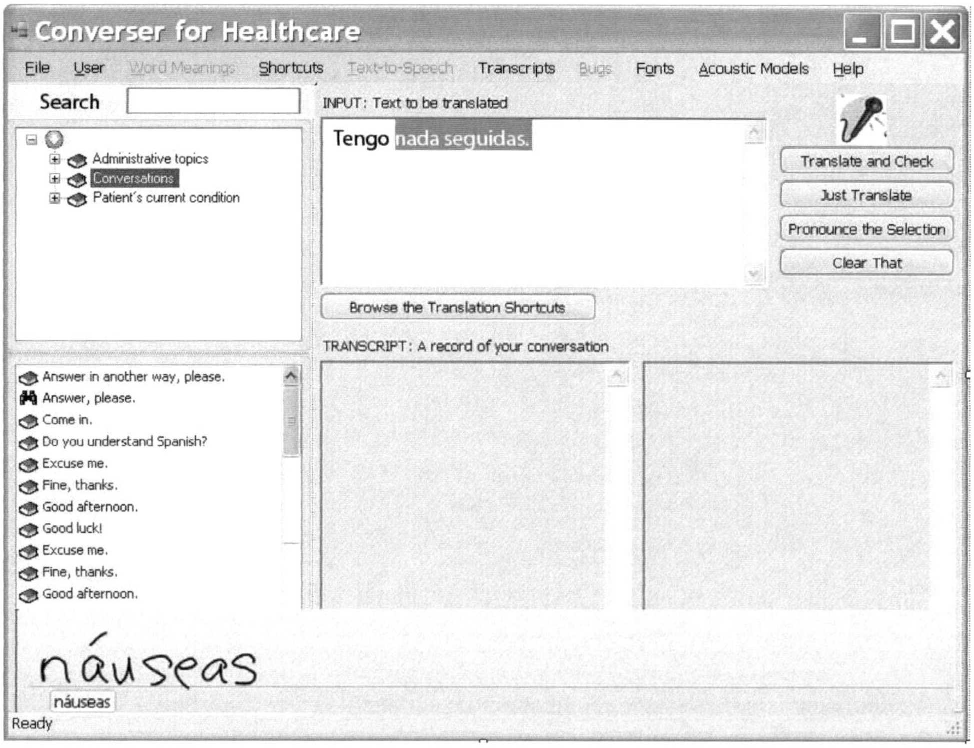

Figure 3: The Input Screen, showing correction of dictation with handwritten input.

Evaluating Task Performance for a Unidirectional Controlled Language Medical Speech Translation System

Nikos Chatzichrisafis, Pierrette Bouillon, Manny Rayner, Marianne Santaholma, Marianne Starlander
University of Geneva, TIM/ISSCO
40 bvd du Pont-d'Arve, CH-1211 Geneva 4, Switzerland

Nikos.Chatzichrisafis@vozZup.com, Pierrette.Bouillon@issco.unige.ch, Emmanuel.Rayner@issco.unige.ch, Marianne.Santaholma@eti.unige.ch, Marianne.Starlander@eti.unige.ch

Beth Ann Hockey
UCSC
NASA Ames Research Center
Moffett Field, CA 94035
bahockey@email.arc.nasa.gov

Abstract

We present a task-level evaluation of the French to English version of MedSLT, a medium-vocabulary unidirectional controlled language medical speech translation system designed for doctor-patient diagnosis interviews. Our main goal was to establish task performance levels of novice users and compare them to expert users. Tests were carried out on eight medical students with no previous exposure to the system, with each student using the system for a total of three sessions. By the end of the third session, all the students were able to use the system confidently, with an average task completion time of about 4 minutes.

1 Introduction

Medical applications have emerged as one of the most promising application areas for spoken language translation, but there is still little agreement about the question of architectures. There are in particular two architectural dimensions which we will address: general processing strategy (statistical or grammar-based), and top-level translation functionality (unidirectional or bidirectional translation). Given the current state of the art in recognition and machine translation technology, what is the most appropriate combination of choices along these two dimensions?

Reflecting current trends, a common approach for speech translation systems is the statistical one. Statistical translation systems rely on parallel corpora of source and target language texts, from which a translation model is trained. However, this is not necessarily the best alternative in safety-critical medical applications. Anecdotally, many doctors express reluctance to trust a translation device whose output is not readily predictable, and most of the speech translation systems which have reached the stage of field testing rely on various types of grammar-based recognition and rule-based translation (Phraselator, 2006; S-MINDS, 2006; MedBridge, 2006). Even though statistical systems exhibit many desirable properties (purely data-driven, domain independence), grammar-based systems utilizing probabilistic context-free grammar tuning appear to deliver better results when training data is sparse (Rayner et al., 2005a).

One drawback of grammar-based systems is that out-of-coverage utterances will be neither recognized nor translated, an objection that critics have sometimes painted as decisive. It is by no means obvious, however, that restricted coverage is such a serious problem. In text processing, work on several generations of controlled language systems has developed a range of techniques for keeping users within the bounds of system coverage (Kittredge, 2003; Mitamura, 1999). If these techniques work for text processing, it is surely not inconceivable that variants of them will be equally successful for spoken language applications. Users are usually able to adapt to a controlled language system given enough time. The critical questions are how to provide efficient support to guide them towards the system's coverage, and how much time they will then need before they have acclimatized.

With regard to top-level translation functionality, the choice is between unidirectional and bidirectional systems. Bidirectional systems are certainly possible today[1], but the arguments in favor of them are not as clear-cut as might first appear. Ceteris paribus, doctors would certainly prefer bidirectional systems; in particular, medical students are trained to conduct examination dialogues using “open questions” (WH-questions), and to avoid leading the patient by asking YN-questions.

The problem with a bidirectional system is, however, that open questions only really work well if the system can reliably handle a broad spectrum of replies from the patients, which is over-optimistic given the current state of the art. In practice, the system's coverage is always more or less restricted, and some experimentation is required before the user can understand what language it is capable of handling. A doctor, who uses the system regularly, will acquire the necessary familiarity. The same might be true for a few patients, if special circumstances mean that they encounter speech translation applications reasonably frequently. Most patients, however, will have had no previous exposure to the system, and may be unwilling to use a type of technology which they have trouble understanding.

A unidirectional system, in which the doctor mostly asks YN-questions, will never be ideal. If, however, the doctor can become proficient in using it, it may still be very much better than the alternative of no translation assistance at all.

To summarize, today’s technology definitely lets us build unidirectional grammar-based medical speech translation systems which work for regular users who have had time to adapt to their limitations. While bidirectional systems are possible, the case for them is less obvious, since users on the patient side may not in practice be able to use them effectively.

In this paper, we will empirically investigate the ability of medical students to adapt to the coverage of unidirectional spoken language translation system. We report a series of experiments, carried out using a French to English speech translation system, in which medical students with no previous experience to the system were asked to use it to carry out a series of verbal examinations on subjects who were simulating the symptoms of various types of medical conditions. Evaluation will be focused on usability. We primarily want to know how quickly subjects learn to use the system, and how their performance compares to that of expert users.

2 The MedSLT system

MedSLT (MedSLT, 2005; Bouillon et al., 2005) is a unidirectional, grammar-based medical speech translation system intended for use in doctor-patient diagnosis dialogues. The system is built on top of Regulus (Regulus, 2006), an Open Source platform for developing grammar-based speech applications. Regulus supports rapid construction of complex grammar-based language models using an example-based method (Rayner et al., 2003; Rayner et al., 2006), which extracts most of the structure of the model from a general linguistically motivated resource grammar. Regulus-based recognizers are reasonably easy to maintain, and grammar structure is shared automatically across different subdomains. Resource grammars are now available for several languages, including English, Japanese (Rayner et al., 2005b), French (Bouillon et al., 2006) and Spanish.

MedSLT includes a help module, whose purpose is to add robustness to the system and guide the user towards the supported coverage. The help module uses a second backup recognizer, equipped with a statistical language model; it matches the

[1] For example, the S-MINDS system (S-MINDS, 2006) offers bidirectional translation.

results from this second recognizer against a corpus of utterances, which are within system coverage and have already been judged to give correct translations. In previous studies (Rayner et al., 2005a; Starlander et al., 2005), we showed that the grammar-based recognizer performs much better than the statistical one on in-coverage utterances, and rather worse on out-of-coverage ones. We also found that having the help module available approximately doubled the speed at which subjects learned to use the system, measured as the average difference in semantic error rate between the results for their first quarter-session and their last quarter-session. It is also possible to recover from recognition errors by selecting one of the displayed help sentences; in the cited studies, we found that this increased the number of acceptably processed utterances by about 10%.

The version of MedSLT used for the experiments described in the present paper was configured to translate from spoken French into spoken English in the headache subdomain. Coverage is based on standard headache-related examination questions obtained from a doctor, and consists mostly of yes/no questions. WH-questions and elliptical constructions are also supported. A typical short session with MedSLT might be as follows:

- is the pain in the side of the head?
- does the pain radiate to the neck?
- to the jaw?
- do you usually have headaches in the morning ?

The recognizer's vocabulary is about 1000 surface words; on in-grammar material, Word Error Rate is about 8% and semantic error rate (per utterance) about 10% (Bouillon et al., 2006). Both the main grammar-based recognizer and the statistical recognizer used by the help system were trained from the same corpus of about 975 utterances. Help sentences were also taken from this corpus.

3 Experimental Setup

In previous work, we have shown how to build a robust and extendable speech translation system. We have focused on performance metrics defined in terms of recognition and translation quality, and tested the system on naïve users without any medical background (Bouillon et al., 2005; Rayner et al., 2005a; Starlander et al., 2005).

In this paper, our primary goal was rather to focus on task performance evaluation using plausible potential users. The basic methodology used is common in evaluating usability in software systems in general, and spoken language systems in particular (Cohen et. al 2000). We defined a simulated situation, where a French-speaking doctor was required to carry out a verbal examination of an English-speaking patient who claimed to be suffering from a headache, using the MedSLT system to translate all their questions. The patients were played by members of the development team, who had been trained to answer questions consistently with the symptoms of different medical conditions which could cause headaches. We recruited eight native French-speaking medical students to play the part of the doctor. All of the students had completed at least four years of medical school; five of them were already familiar with the symptoms of different types of headaches, and were experienced in real diagnosis situations.

The experiment was designed to study how well users were able to perform the task using the MedSLT system. In particular, we wished to determine how quickly they could adapt to the restricted language and limited coverage of the system. As a comparison point, representing near-perfect performance, we also carried out the same test on two developers who had been active in implementing the system, and were familiar with its coverage.

Since it seemed reasonable to assume that most users would not interact with the system on a daily basis, we conducted testing in three sessions, with an interval of two days between each session. At the beginning of the first session, subjects were given a standardized 10-minute introduction to the system. This consisted of instruction on how to set up the microphone, a detailed description of the MedSLT push-to-talk interface, and a video clip showing the system in action. At the end of the presentation, the subject was given four sample sentences to get familiar with the system.

After the training was completed, subjects were asked to play the part of a doctor, and conduct an examination through the system. Their task was to identify the headache-related condition simulated by the "patient", out of nine possible conditions. Subjects were given definitions of the simulated headache types, which included conceptual information about location, duration, frequency, onset and possible other symptoms the particular type of headache might exhibit.

Subjects were instructed to signal the conclusion of their examination when they were sure about the type of simulated headache. The time required to reach a conclusion was noted in the experiment protocols by the experiment supervisor.

The subjects repeated the same diagnosis task on different predetermined sets of simulated conditions during the second and third sessions. The sessions were concluded either when a time limit of 30 minutes was reached, or when the subject completed three headache diagnoses. At the end of the third session, the subject was asked to fill out a questionnaire.

4 Results

Performance of a speech translation system is best evaluated by looking at system performance as a whole, and not separately for each subcomponent in the systems processing pipeline (Rayner et. al. 2000, pp. 297-pp. 312). In this paper, we consequently focus our analysis on objective and subjective usability-oriented measures.

In Section 4.1, we present objective usability measures obtained by analyzing user-system interactions and measuring task performance. In Section 4.2, we present subjective usability figures and a preliminary analysis of translation quality.

4.1 Objective Usability Figures

4.1.1 Analysis of User Interactions

Most of our analysis is based on data from the MedSLT system log, which records all interactions between the user and the system. An interaction is initiated when the user presses the "Start Recognition" button. The system then attempts to recognize what the user says. If it can do so, it next attempts to show the user how it has interpreted the recognition result, by first translating it into the Interlingua, and then translating it back into the source language (in this case, French). If the user decides that the back-translation is correct, they press the "Translate" button. This results in the system attempting to translate the Interlingua representation into the target language (in this case, English), and speak it using a Text-To-Speech engine. The system also displays a list of "help sentences", consisting of examples that are known to be within coverage, and which approximately match the result of performing recognition with the statistical language model. The user has the option of choosing a help sentence from the list, using the mouse, and submitting this to translation instead.

We classify each interaction as either "successful" or "unsuccessful". An interaction is defined to be unsuccessful if either

i) the user re-initiates recognition without asking the system for a translation, or

ii) the system fails to produce a correct translation or back translation.

Our definition of "unsuccessful interaction" includes instances where users accidentally press the wrong button (i.e. "Start Recognition" instead of "Translate"), press the button and then say nothing, or press the button and change their minds about what they want to ask half way through. We observed all of these behaviors during the tests.

Interactions where the system produced a translation were counted as successful, irrespective of whether the translation came directly from the user's spoken input or from the help list. In at least some examples, we found that when the translation came from a help sentence it did not correspond directly to the sentence the user had spoken; to our surprise, it could even be the case that the help sentence expressed the directly opposite question to the one the user had actually asked. This type of interaction was usually caused by some deficiency in the system, normally bad recognition or missing coverage. Our informal observation, however, was that, when this kind of thing happened, the user perceived the help module positively: it enabled them to elicit at least some information from the patient, and was less frustrating than being forced to ask the question again.

Table I to Table III show the number of total interactions per session, the proportion of successful interactions, and the proportion of interactions completed by selecting a sentence from the help list. The total number of interactions required to complete a session decreased over the three sessions, declining from an average of 98.6 interactions in the first session to 63.4 in the second (36% relative) and 53.9 in the third (45% relative). It is interesting to note that interactions involving the help system did not decrease in frequency, but remained almost constant over the first two sessions

(15.5% and 14.0%), and were in fact most common during the third session (21.7%).

Session 1			
Subject	**Interactions**	**% Successful**	**% Help**
User 1	57	56.1%	0.0%
User 2	98	52.0%	25.5%
User 3	91	63.7%	15.4%
User 4	156	69.9%	10.3%
User 5	86	64.0%	22.1%
User 6	134	47.0%	19.4%
User 7	56	53.6%	5.4%
User 8	111	63.1%	26.1%
AVG	98.6	58.7%	15.5%

Table I Total interaction rounds, percentage of successful interactions, and interactions involving the help system by subject for the 1st session

Session 2			
Subject	**Interactions**	**% Successful**	**% Help**
User 1	50	74.0%	2.0%
User 2	63	55.6%	27.0%
User 3	34	88.2%	23.5%
User 4	96	57.3%	17.7%
User 5	64	65.6%	21.9%
User 6	93	68.8%	10.8%
User 7	48	60.4%	4.2%
User 8	59	79.7%	5.1%
AVG	*63.4*	*68.7%*	*14.0%*

Table II Total interaction rounds, percentage of successful interactions, and interactions involving the help system by subject for the 2nd session

Session 3			
Subject	**Interactions**	**% Successful**	**% Help**
User 1	33	90.9%	33.3%
User 2	57	56.1%	22.8%
User 3	48	72.9%	29.2%
User 4	67	70.2%	16.4%
User 5	68	73.5%	27.9%
User 6	60	70.0%	6.7%
User 7	41	65.9%	14.6%
User 8	57	56.1%	22.8%
AVG	*53.9*	*69.5%*	*21.7%*

Table III Total interaction rounds, percentage of successful interactions, and interactions involving the help system by subject for the 3rd session

In order to establish a performance baseline, we also analyzed interaction data for two expert users, who performed the same experiment. The expert users were two native French-speaking system developers, which were both familiar with the diagnosis domain. Table IV summarizes the results of those users. One of our expert users, listed as Expert 2, is the French grammar developer, and had no failed interactions. This confirms that recognition is very accurate for users who know the coverage.

Session 1 / Expert Users			
Subject	**Interactions**	**% Successful**	**% Help**
Expert 1	36	77.8%	13.9%
Expert 2	30	100.0%	3.3%
AVG	*33*	*88.9%*	*8.6%*

Table IV Number of interactions, and percentages of successful interactions, and interactions involving the help component

The expert users were able to complete the experiment using an average of 33 interaction rounds. Similar performance levels were achieved by some subjects during the second and third session, which suggests that it is possible for at least some new users to achieve performance close to expert level within a few sessions.

4.1.2 Task Level Performance

One of the important performance indicators for end users is how long it takes to perform a given task. During the experiments, the instructors noted completion times required to reach a definite diagnosis in the experiment log. Table VI shows task completion times, categorized by session (columns) and task within the session (rows).

	Session 1	**Session 2**	**Session 3**
Diagnosis 1	17:00 min	11:00 min	7:54 min
Diagnosis 2	11:00 min	6:18 min	5:34 min
Diagnosis 3	7:54 min	4:10 min	4:00 min

Table V Average time required by subjects to complete diagnoses

In the last two sessions, after subjects had acclimatized to the system, a diagnosis takes an average of about four minutes to complete. This compares to a three-minute average required to complete a diagnosis by our expert users.

4.1.3 System coverage

Table VI shows the percentage of in-coverage sentences uttered by the users on interactions that did not involve invocation of the help component.

	IN-COVERAGE SENTENCES
Session 1	54.9%
Session 2	60.7%
Session 3	64.6%

Table VI Percentage of in-coverage sentences

This indicates that subjects learn and adapt to the system coverage as they use the system more. The average proportion of in-coverage utterances is 10 percent higher during the third session than during the first session.

4.2 Subjective Usability Measures

4.2.1 Results of Questionnaire

After finishing the third session, subjects were asked to fill in a short questionnaire, where responses were on a five-point scale ranging from 1 ("strongly disagree") to 5 ("strongly agree"). The results are presented in Table VIII.

STATEMENT	SCORE
I quickly learned how to use the system.	4.4
System response times were generally satisfactory.	4.5
When the system did not understand me, the help system usually showed me another way to ask the question.	4.6
When I knew what I could say, the system usually recognized me correctly.	4.3
I was often unable to ask the questions I wanted.	3.8
I could ask enough questions that I was sure of my diagnosis.	4.3
This system is more effective than non-verbal communication using gestures.	4.3
I would use this system again in a similar situation.	4.1

Table VIII Subject responses to questionnaire. Scores are on a 5-point scale, averaged over all answers.

Answers were in general positive, and most of the subjects were clearly very comfortable with the system after just an hour and a half of use. Interestingly, even though most of the subjects answered "yes" to the question "I was often unable to ask the questions I wanted", the good performance of the help system appeared to compensate adequately for missing coverage.

4.2.2 Translation Performance

In order to evaluate the translation quality of the newly developed French-to-English system, we conducted a preliminary performance evaluation, similar to the evaluation method described in (Bouillon 2005).

We performed translation judgment in two rounds. In the first round, an English-speaking judge was asked to categorize target utterances as comprehensible or not without looking at corresponding source sentences. 91.1% of the sentences were judged as comprehensible. The remaining 8.9% consisted of sentences where the terminology used was not familiar to the judge and of sentences where the translation component failed to produce a sufficiently good translation. An example sentence is

- Are the headaches better when you experience dark room?

which stems from the French source sentence

- Vos maux de tête sont ils soulagés par obscurité?

In the second round, English-speaking judges, sufficiently fluent in French to understand source language utterances, were shown the French source utterance, and asked to decide whether the target language utterance correctly reflected the meaning of the source language utterance. They were also asked to judge the style of the target language utterance. Specifically, judges were asked to classify sentences as "BAD" if the meaning of the English sentence did not reflect the meaning of the French sentence. Sentences were categorized as "OK" if the meaning was transferred correctly and the sentence was comprehensible, but the style of the resulting English sentence was not perfect. Sentences were judged as "GOOD" when they were comprehensible, and both meaning and style were considered to be completely correct. Table VIII summarizes results of two judges.

	Good	OK	Bad
Judge 1	15.8%	73.80%	10.3%
Judge 2	46.6%	47.1%	6.3%

Table VIII Judgments of the quality of the translations of 546 utterances

It is apparent that translation judging is a highly subjective process. When translations were marked as "bad", the problem most often seemed to be related to lexical items where it was challenging to find an exact correspondence between French and English. Two common examples were "troubles de la vision", which was translated as "blurred vision", and "faiblesse musculaire", which was translated as "weakness". It is likely that a more careful choice of lexical translation rules would deal with at least some of these cases.

5 Summary

We have presented a first end-to-end evaluation of the MedSLT spoken language translation system. The medical students who tested it were all able to use the system well, with performance in some cases comparable to that of that of system developers after only two sessions. At least for the fairly simple type of diagnoses covered by our scenario, the system's performance appeared clearly adequate for the task.

This is particularly encouraging, since the French to English version of the system is quite new, and has not yet received the level of attention required for a clinical system. The robustness added by the help system was sufficient to compensate for that, and in most cases, subjects were able to find ways to maneuver around coverage holes and other problems. It is entirely reasonable to hope that performance, which is already fairly good, would be substantially better with another couple of months of development work.

In summary, we feel that this study shows that the conservative architecture we have chosen shows genuine potential for use in medical diagnosis situations. Before the end of 2006, we hope to have advanced to the stage where we can start initial trials with real doctors and patients.

Acknowledgments

We would like to thank Agnes Lisowska, Alia Rahal, and Nancy Underwood for being impartial judges over our system's results.

This work was funded by the Swiss National Science Foundation.

References

P. Bouillon, M. Rayner, N. Chatzichrisafis, B.A. Hockey, M. Santaholma, M. Starlander, Y. Nakao, K. Kanzaki, and H. Isahara. 2005. *A generic multilingual open source platform for limited-domain medical speech translation.* In Proceedings of the 10th Conference of the European Association for Machine Translation (EAMT), Budapest, Hungary.

P. Bouillon, M. Rayner, B. Novellas, Y. Nakao, M. Santaholma, M. Starlander, and N. Chatzichrisafis. 2006. *Une grammaire multilingue partagée pour la reconnaissance et la génération.* In Proceedings of TALN 2006, Leuwen, Belgium.

M. Cohen, J. Giangola, and J. Balogh. 2004, Voice User Interface Design. Addison Wesley Publishing.

R. I. Kittredge. 2003. *Sublanguages and comtrolled languages.* In R. Mitkov, editor, The Oxford Handbook of Computational Linguistics, pages 430–447. Oxford University Press.

MedBridge, 2006. *http://www.medtablet.com/.* As of 15th March 2006.

MedSLT, 2005. http://sourceforge.net/projects/medslt/. As of 15th March 2006.

T. Mitamura. 1999. *Controlled language for multilingual machine translation.* In Proceedings of Machine Translation Summit VII, Singapore.

Phraselator, 2006. *http://www.phraselator.com.* As of 15 February 2006.

M. Rayner, B.A. Hockey, and J. Dowding. 2003. *An open source environment for compiling typed unification grammars into speech recognisers.* In Proceedings of the 10th EACL (demo track), Budapest, Hungary.

M. Rayner, N. Chatzichrisafis, P. Bouillon, Y. Nakao, H. Isahara, K. Kanzaki, and B.A. Hockey. 2005b. *Japanese speech understanding using grammar specialization.* In HLT-NAACL 2005: Demo Session, Vancouver, British Columbia, Canada. Association for Computational Linguistics.

M. Rayner, P. Bouillon, N. Chatzichrisafis, B.A. Hockey, M. Santaholma,M. Starlander, H. Isahara,

K. Kankazi, and Y. Nakao. 2005a. *A methodology for comparing grammar-based and robust approaches to speech understanding*. In Proceedings of the 9th International Conference on Spoken Language Processing (ICSLP), Lisboa, Portugal.

M. Rayner, D. Carter, P. Bouillon, V. Digalakis, and M. Wirén. 2000. The Spoken Language Translator, Cambridge University Press.

M. Rayner, N. Chatzichrisafis, P. Bouillon, Y. Nakao, H. Isahara, K. Kanzaki, and B.A. Hockey. 2005b. *Japanese speech understanding using grammar specialization*. In HLT-NAACL 2005: Demo Session, Vancouver, British Columbia, Canada. Association for Computational Linguistics.

M. Rayner, B.A. Hockey, and P. Bouillon. 2006. *Putting Linguistics into Speech Recognition: The Regulus Grammar Compiler*. CSLI Press, Chicago.

Regulus, 2006. *http://sourceforge.net/projects/regulus/*. As of 15 March 2006.

S-MINDS, 2006. *http://www.sehda.com/*. As of 15 March 2006.

M. Starlander, P. Bouillon, N. Chatzichrisafis, M. Santaholma, M. Rayner, B.A. Hockey, H. Isahara, K. Kanzaki, and Y. Nakao. 2005. *Practicing controlled language through a help system integrated into the medical speech translation system (MedSLT)*. In Proceedings of the MT Summit X, Phuket, Thailand

Speech to Speech Translation for Medical Triage in Korean

Farzad Ehsani, Jim Kimzey, Demitrios Master, Karen Sudre
Engineering Department
Sehda, Inc.
Mountain View, CA 94043
{farzad,jkimzey,dlm,karen}@sehda.com

Hunil Park
Independent Consultant
Seoul, Korea
phunil@hotmail.com

Abstract

S-MINDS is a speech translation engine, which allows an English speaker to communicate with a non-English speaker easily within a question-and-answer, interview-style format. It can handle limited dialogs such as medical triage or hospital admissions. We have built and tested an English-Korean system for doing medical triage with a translation accuracy of 79.8% (for English) and 78.3% (for Korean) for all non-rejected utterances. We will give an overview of the system building process and the quantitative and qualitatively system performance.

1 Introduction

Speech translation technology has the potential to give nurses and other clinicians immediate access to consistent, easy-to-use, and accurate medical interpretation for routine patient encounters. This could improve safety and quality of care for patients who speak a different language from that of the healthcare provider.

This paper describes the building and testing of a speech translation system, S-MINDS (Speaking Multilingual Interactive Natural Dialog System), built in less than 4 months from specification to the test scenario described. Although this paper shows a number of areas for improvement in the S-MINDS system, it does demonstrate that building and deploying a successful speech translation system is becoming possible and perhaps even commercially viable.

2 Background

Sehda is focused on creating speech translation systems to overcome language barriers in healthcare settings in the U.S. The number of people in the U.S. who speak a language other than English is large and growing, and Spanish is the most commonly spoken language next to English. According to the 2000 census, 18% of the U.S. population aged 5 and older (47 million people) did not speak English at home.[1] This represents a 48% increase from the 1990 figure. In 2000, 8% of the population (21 million) was Limited English Proficient (LEP). More than 65% of the LEP population (almost 14 million people) spoke Spanish.

A body of research shows that language barriers impede access to care, compromise quality, and increase the risk of adverse outcomes. Although trained medical interpreters and bilingual healthcare providers are effective in overcoming such language barriers, the use of semi-fluent healthcare professionals and ad hoc interpreters causes more interpreter errors and lower quality of care (Flores 2005).

One study analyzed the problem of language barriers for hospitalized inpatients. The study, which focused on pediatric patients, sought to determine whether patients whose families have a language barrier are more likely to incur serious medical errors than patients without a language barrier (Cohen et al., 2005). The study's conclusion was that patients of LEP families had a twofold increased risk for serious medical incident compared with patients whose families did not have a language barrier. It is important to note that the LEP

1 US Census Bureau, 2000

patients in this study were identified as needing interpreters during their inpatient stay and medical interpreters were available.

Although the evidence favors using trained medical interpreters, there is a gap between best practice and reality. Many patients needing an interpreter do not get one, and many must use ad hoc interpreters. In a study of 4,161 uninsured patients who received care in 23 hospitals in 16 cities, more than 50% who needed an interpreter did not get one (Andrulis et al., 2002).

Another study surveyed 59 residents in a pediatric residency program in an urban children's hospital (O'Leary and Hampers, 2003). Forty of the 59 residents surveyed spoke little or no Spanish. Again, it is important to note that this hospital had in-house medical interpreters. Of this group of nonproficient residents:

- 100% agreed that the hospital interpreters were effective; however, 75% "never" or only "sometimes" used the hospital interpreters.
- 53% used their inadequate language skills in the care of patients "often" or "every day."
- 53% believed the families "never" or only "sometimes" understood their child's diagnosis.
- 43% believed the families "never" or only "sometimes" understood discharge instructions.
- 40% believed the families "never" or only "sometimes" understood the follow-up plan.
- 28% believed the families "never" or only "sometimes" understood the medications.
- 53% reported calling on their Spanish-proficient colleagues "often" or "every day" for help.
- 80% admitted to avoiding communication with non-English-speaking families.

The conclusion of the study was as follows: "Despite a perception that they are providing suboptimal communication, nonproficient residents rarely use professional interpreters. Instead, they tend to rely on their own inadequate language skills, impose on their proficient colleagues, or avoid communication with Spanish-speaking families with LEP."

Virtually every study on language barriers suggests that these residents are not unique. Physicians and staff at several hospitals have told Sehda that they are less likely to use a medical interpreter or telephone-based interpreter because it takes too long and is too inconvenient. Sehda believes that to bridge this gap requires 2-way speech translation solutions that are immediately available, easy to use, accurate, and consistent in interpretation.

The need for speech translation exists in healthcare, and a lot of work has been done in speech translation over the past two decades. Carnegie-Mellon University has been experimenting with spoken language translation in its JANUS project since the late 1980s (Waibel et al., 1996). The University of Karlsruhe, Germany, has also been involved in an expansion of JANUS. In 1992, these groups joined ATR in the C-STAR consortium (Consortium for Speech Translation Advanced Research) and in January 1993 gave a successful public demonstration of telephone translation between English, German and Japanese, within the limited domain of conference registrations (Woszczyna, 1993). A number of other large companies and laboratories including NEC (Isotani, et al., 2003) in Japan, the Verbmobil Consortium (Wahlster, 2000), NESPOLE! Consortium (Florian et al., 2002), AT&T (Bangalore and Riccardi, 2001), and ATR have been making their own research effort (Yasuda et al., 2003). LC-Star and TC-Star are two recent European efforts to gather the data and the industrial requirements to enable pervasive speech-to-speech translation (Zhang, 2003). Most recently, the DARPA TransTac program (previously known as Babylon) has been focusing on developing deployable systems for English to Iraqi Arabic.

3 System Description

Unlike other systems that try to solve the speech translation problem with the assumption that there is a moderate amount of data available, S-MINDS focuses on rapid building and deployment of speech translation systems in languages where little or no data is available. S-MINDS allows the user to communicate easily in a question-and-answer, interview-style conversation across languages in limited domains such as border control, hospital admissions or medical triage, or other narrow interview fields.

S-MINDS uses a number of voice-independent speech recognition engines with the usage dependent on the languages and the particular domain. These engines include Nuance 8.5[2], SRI EduSpeak 2.0[3], and Entropic's HTK-based engine.[4] There is a dialog/translation creation tool that allows us to compile and run our created dialogs with any of these engines. This allows our developers to be free from the nuances of any particular engine that is deployed. S-MINDS uses a combination of grammars and language models with these engines, depending on the task and the availability of training data. In the case of the system described in this document, we were using Nuance 8.5 for both English and Korean speech recognition.

We use our own semantic parser, which identifies keywords and phrases that are tagged by the user; these in turn are fed into an interpretation engine. Because of the limited context, we can achieve high translation accuracy with the interpretation engine. However, as the name suggests, this engine does not directly translate users' utterances but interprets what they say and paraphrases their statements. Finally, we use a voice generation system (which splices human recordings) along with the Festival TTS engine to output the translations. This has been recently replaced by the Cepstral TTS engine.

Additionally, S-MINDS includes a set of tools to modify and augment the existing system with additional words and phrases in the field in a matter of a few minutes.

The initial task given to us was a medical disaster recovery scenario that might occur near an American military base in Korea. We were given about 270 questions and an additional 90 statements that might occur on the interviewer side. Since our system is an interview-driven system (sometimes referred to as "1.5-way"), the second-language person is not given the option of initiating conversations. The questions and statements given to us covered several domains related to the task above, including medical triage, force protection at the installation gate, and some disaster recovery questions. In addition to the 270 assigned questions, we created 120 of our own in order to make the domains more complete.

3.1 Data Collection

Since we assumed that we could internally generate the English language data used to ask the question but not the language data on the Korean side, our entire focus for the data collection task was on Korean. As such, we collected about 56,000 utterances from 144 people to answer the 390 questions described above. This data collection was conducted over the course of 2 months via a telephone-based computer system that the native Korean speakers could call. The system first introduced the purpose of the data collection and then presented the participants with 12 different scenarios. The participants were then asked a subset of the questions after each of the scenarios. One advantage of the phone-based system – in addition to the savings in administrative costs – was that the participants were free to do the data collection any time during the day or night, from any location. The system also allowed participants to hang up and call back at a later time. The participants were paid only if they completed all the scenarios.

Of this data, roughly 7% was unusable and was thrown away. Another 31% consisted of one-word answers (like "yes"). The rest of the data consisted of utterances 2 to 25 words long. Approximately 85% of the usable data was used for training; the remainder was used for testing.

The transcription of the data started one week after the start of the data collection, and we started building the grammars three weeks later.

3.2 System Development

We have an extensive set of tools that allow nonspecialists, with a few days of training, to build complete mission-oriented domains. In this project, we used three bilingual college graduates who had no knowledge of linguistics. We spent the first 10 days training them and the next two weeks closely supervising their work. Their work involved taking the sentences that were produced from the data collection and building grammars for them until the "coverage" of our grammars – that is, the num-

[2] http://www.nuance.com/nuancerecognition/

[3] http://www.speechatsri.com/products/eduspeak.shtml

[4] http://htk.eng.cam.ac.uk/

ber of utterances from the training set that our system would handle – was larger than a set threshold (generally set between 80% and 90%). Because of the scarcity of Korean-language data, we built this system based entirely on grammar language models rather than statistical language models. Grammars are generally more rigid than statistical language models, and as such grammars tend to have higher in-domain accuracy and much lower out-of-domain accuracy[5] than statistical language models. This means that the system performance will depend greatly upon on how well our grammars cover the domains.

The semantic tagging and the paraphrase translations were built simultaneously with the grammars. This involved finding and tagging the semantic classes as well as the key concepts in each utterance. Frame-based translations were performed by doing concept and semantic transfer. Because our tools allowed the developers to see the resulting frame translations right away, they were able to make fixes to the system as they were building it; hence, the system-building time was greatly reduced.

We used about 15% of the collected telephone data for batch testing. Before deployment, our average word accuracy on the batch results was 92.9%. The translation results were harder to measure directly, mostly because of time constraints.

3.3 System Testing

We tested our system with 11 native Korean speakers, gathering 968 utterances from them. The results of the test are shown in Table 1. Most of the valid rejected utterances occurred because participants spoke too softly, too loudly, before the prompt, or in English. Note that there was one utterance with bad translation; that and a number of other problems were fixed before the actual field testing.

5 Note that there are many factors effecting both grammar-based and statistical language model based speech recognition, including noise, word perplexity, acoustic confusability, etc. The statement above has been true with some of the experiments that we have done, but we can not claim that it is universally true.

Category	Percentage
Total Recognized Correctly	82.0%
Total Recognized Incorrectly	5.8%
Total Valid Rejection	8.0%
Total Invalid Rejected	4.1%
Total unclear translations	0.1%

Table 1: Korean-to-English system testing results for the 11 native Korean speakers.

4 Experimental Setup

A military medical group used S-MINDS during a medical training exercise in January 2005 in Carlsbad, California. The testing of speech translation systems was integrated into the exercise to assess the viability of such systems in realistic situations. The scenario involved a medical aid station near the front lines treating badly injured civilians. The medical facilities were designed to quickly triage severely wounded patients, provide life-saving surgery if necessary, and transfer the patients to a safer area as soon as possible.

4.1 User Training

Often the success or failure of these interactive systems is determined by how well the users are trained on the systems' features.

Training and testing on S-MINDS took place from November 2004 through January 2005. The training had three parts: a system demonstration in November, two to three hours of training per person in December, and another three-hour training session in January. About 30 soldiers were exposed to S-MINDS during this period. Because of the tsunami in Southeast Asia, many of the people who attended the November demo and December training were not available for the January training and the exercise. Nine service members used S-MINDS during the exercise. Most of them had attended only the training session in January.

4.2 Test Scenarios

Korean-speaking 'patients' arrived by military ambulance. They were received into one of three tents where they were (notionally) triaged, treated, and prepared for surgery. The tents were about 20 feet wide by 25 feet deep, and each had six to eight cots for patients. The tents had lights and electricity.

The environment was noisy, sandy, and 'bloody.' The patients' makeup coated our handsets by the end of the day. There were many soldiers available to help and watch. Nine service members used S-MINDS during a four-hour period.

All of the 'patients' spoke both English and Korean. A few 'patients' were native Korean speakers, and two were American service members who spoke Korean fairly fluently but with an accent. The 'patients' were all presented as severely injured from burns, explosions, and cuts and in need of immediate trauma care.

The 'patients' were instructed to act as if they were in great pain. Some did, and they sounded quite realistic. In fact, their recorded answers to questions were sometimes hard for a native Korean speaker to understand. The background noise in the tents was quite loud (because of the number of people involved, screaming patients and close quarters). Although we did not directly measure the noise; we estimate it ranged from 65 to 75 decibels.

4.3 Physical and Hardware Setup

S-MINDS is a flexible system that can be configured in different ways depending on the needs of the end user. Because of the limited time available for training, the users were trained on a single hardware setup, tailored to our understanding of how the exercises would be conducted. Diagrams available before the exercises showed that each tent would have a "translation station" where Korean-speaking patients would be brought. The experimenters (two of the authors) had expected that the tents would be positioned at least 40 feet apart. In reality, the tents were positioned about 5 feet apart, and there was no translation station.

Our original intent was to use S-MINDS on a Sony U-50 tablet computer mounted on a computer stand with a keyboard and mouse at the translation station, and for a prototype wireless device – based on a Bluetooth-like technology to eliminate the need for wires between the patient and the system – that we had built previously. However, because of changes in the conduct of the exercise, the experimenters had to step in and quickly set up two of the S-MINDS systems without the wireless system (because of the close proximity of the tents) and without the computer stands. The keyboards and mice were also removed so that the S-MINDS systems could be made portable. The medics worked in teams of two; one medic would hold the computer and headset for the injured patient while the other medic conducted the interview.

5 Results

The nine participants used our system to communicate with 'patients' over a four-hour period. We analyzed qualitative problems with using the system and quantitative results of translation accuracy.

5.1 Problems with System Usage

We observed a number of problems in the test scenarios with our system. These represent some of the more common problems with the S-MINDS system. The authors suspect these may be endemic of all such systems.

5.1.1 Inadequate Training on the System

Users were trained to use the wireless units, which interfered with each other when used in close proximity. For the exercise, we had to set up the units without the wireless devices because the users had not been trained on this type of setup. As a result, service members were forced to use a different system from the one they were trained on.

Also, the users had difficulty navigating to the right domain. S-MINDS has multiple domains each optimized for a particular scenario (medical triage, pediatrics, etc.), but the user training did not include navigation among domains.

5.1.2 User Interface Issues

The user interface and the system's user feedback messages caused unnecessary confusion with the interviewers. The biggest problem was that the system responded with, "I'm sorry, I didn't hear that clearly" whenever a particular utterance wasn't recognized. This made the users think they should just repeat their utterance over and over. In fact, the problem was that they were saying something that were out of domain or did not fit any dialogs in S-MINDS, so no matter how many times they repeated the phrase, it would not be recognized. This caused the users significant frustration.

5.2. Quantative Analysis

During the system testing, there were 363 recorded interactions for the English speakers. Unfortunately, the system was not set up to record the utterances that had a very low confidence score (as determined by the Nuance engine), and the user was asked to repeat those utterances again. Here is the rough breakdown for all of the English interactions:

- 52.5% were translated correctly into Korean
- 34.2% were rejected by the system
- 13.3% had misrecognition or mistranslation errors

This means that S-MINDS tried to recognize and translate 65.8% of the English utterances and of those 79.8% were correctly translated. A more detailed analysis is presented in Figure 1.

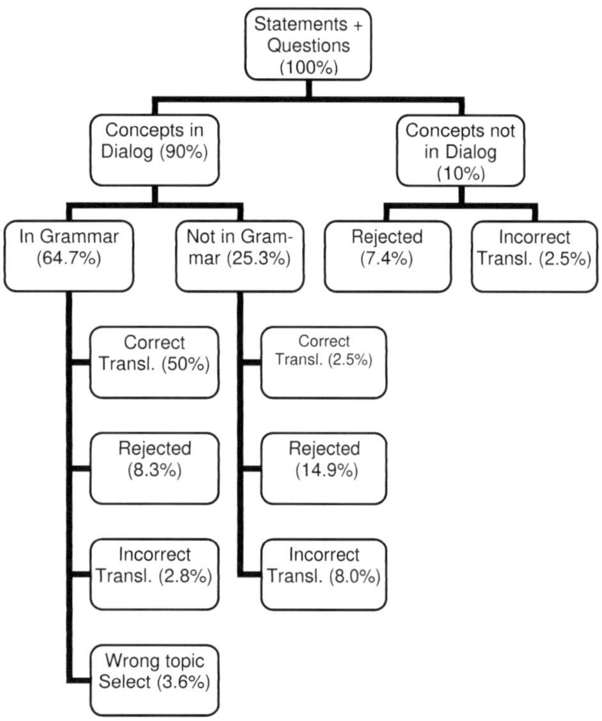

Figure 1: Detailed breakdown for the English utterances and percentage breakdown for each category.

The Korean speakers' responses to each of the questions that were recognized and translated are analyzed in Figure 2. Note that the accuracy for the non-rejected responses is 78.3%.

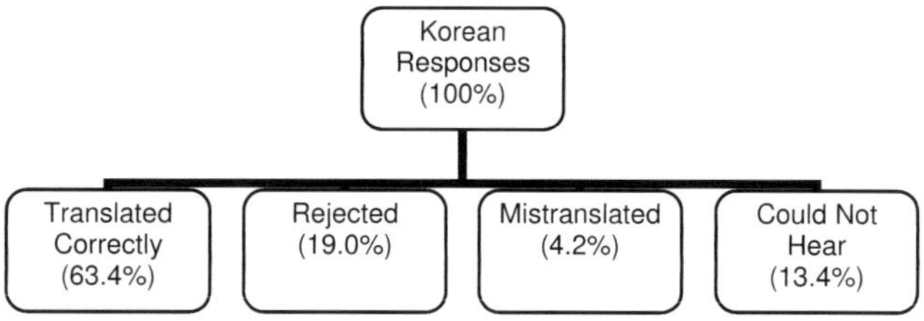

Figure 2: Detailed breakdown of the recognition for the Korean utterances and percentage breakdown for each category.

6 Discussion

Although these results are less than impressive, a close evaluation pointed to three areas where a concentration of effort would significantly improve translation accuracy and reduce mistranslations. These areas were:

1) Data collection with English speakers to increase coverage on the dialogs.
   a) 34% of the things the soldiers said were things S-MINDS was not designed to translate.
   b) We had assumed that our existing English system would have adequate coverage without any additional data collection.
2) User verification on low-confidence results.
3) Improved feedback prompts when a phrase is not recognized; for example:
   a) One user said, "Are you allergic to any allergies?" three times before he caught himself and said, "Are you allergic to any medications?"
   b) Another user said, "How old are you?" seven times before realizing he needed to switch to a different domain, where he was able to have the phrase translated.
   c) Another user repeated, "What is your name?" nine times before giving up on the phrase (this phrase wasn't in the S-MINDS Korean medical mission set).

Beyond improving the coverage, the system's primary problem seemed to be in the voice user interface since even the trained users had a difficult time in using the system.

The attempt at realism in playing out a high-trauma scenario may have detracted from the effectiveness of the event as a test of the systems' abilities under more routine (but still realistic) conditions.

7 New Results

Based on the results of this experiment, we had a secondary deployment in a medical setting for a very similar system.

We applied what we had learned to that setting and achieved better results in a few areas. For example:

1. Data collection in English helped tremendously. S-MINDS recognized about 40% more concepts than it had been able to recognize using only grammars created by subject-matter experts.
2. Verbal verification of the recognized utterance was added to system, and that improved the user confidence, although too much verification tended to frustrate the users.
3. Feedback prompts were designed to give more specific feedback, which seemed to reduce user frustration and the number of mistakes.

Overall, the system performance seemed to improve. We continue to gather data on this task, and we believe that this is going to enable us to identify the next set of problems that need to be solved.

8 Acknowledgement

This research was funded in part by the LASER ACTD. We specially wish to thank Mr. Pete Fisher of ARL for his generous support and his participation in discussions related to this project.

References

Andrulis Dennis, Nanette Goodman, Carol Pryor (2002), "What a Difference an Interpreter Can make" April 2002. Access Project, www.accessproject.org/downloads/c_LEPreport ENG.pdf

Bangalore, S. and G. Riccardi, (2001), "A Finite State Approach to Machine Translation," North American ACL 2001, Pittsburgh.

Cohen, L, F. Rivara, E. K. Marcuse, H. McPhillips, and R. Davis, (2005), "Are Language Barriers Associated With Serious Medical Events in Hospitalized Pediatric Patients?", *Pediatrics*, September 1, 2005; 116(3): 575 - 579

Flores Glenn, (2005), "The Impact of Medical Interpreter Services on the Quality of Health Care: A Systematic Review," *Medical Care Research and Review*, Vol. 62, No. 3, pp. 255-299

Florian M., et. al. (2002), "Enhancing the Usability and Performance of NESPOLE!: a Real-World Speech-to-Speech Translation System", HLT 2002, San Diego, California U.S., March 2002.

Isotani, R., Kiyoshi Yamabana, Shinichi Ando, Ken Hanazawa, Shin-ya Ishikawa and Ken-ichi ISO (2003), "Speech-to-Speech Translation Software on PDAs for Travel Conversation," NEC Research and Development, Apr. 2003, Vol.44, No.2.

O'Leary and Hampers (2003) "The Truth About Language Barriers: One Residency Program's Experience," *Pediatrics*, May 1, 2003; 111(5): pp. 569 - 573.

Keiji Yasuda, Eiichiro Sumita, Seiichi Yamamoto, Genichiro Kikui, Masazo Yanagida, "Real-Time Evaluation Architecture for MT Using Multiple Backward Translations," *Recent Advances in Natural Language Processing*, pp. 518-522, Sep., 2003

Wahlster, W. (2000), *Verbmobil: Foundations of Speech-to-Speech Translation.* Springer.

Waibel, A., (1996), "Interactive Translation of Conversational Speech," *IEEE Computer*, July 1996, 29-7, pp. 41-48.

Woszczyna, et al., (1993), "Recent Advances in JANUS: A Speech Translation System," *DARPA Speech and Natural Language Workshop 1993*, session 6 – MT.

Zhang, Ying, (2003), "Survey of Current Speech Translation Research," Found on Web: http://projectile.is.cs.cmu.edu/research/public/tal ks/ speechTranslation/sst-survey-joy.pdf

Automated Interpretation of Clinical Encounters with Cultural Cues and Electronic Health Record Generation

Daniel T. Heinze, PhD
A-Life Medical, Inc.
6055 Lusk Blvd. – Suite 200
San Diego, CA 92130
dheinze@alifemedical.com

Alexander Turchin, MD, MS
Brigham and Women's Hospital
221 Longwood Ave.
Boston, MA 02115
aturchin@partners.org

V. Jagannathan, PhD
West Virginia University and MedQuist, Inc.
235 High Street, Suite 213
Morgantown, WV 26505
juggy@medquist.com

Abstract

A review of publications by and about medical interpreters reveals a number of operational similarities and shared attitudes and beliefs with the medical coding and abstracting community as it existed ten years ago in the mid-1990's. At that time, the first of what have now become several successful commercial products using Natural Language Processing (NLP) for automated coding and abstracting appeared. The initial reaction was that machines could never do what human coders and abstractors do, and anecdotal accounts illustrating the difficulty of the task proliferated. The claims of superior human capabilities and the accuracy of the anecdotal accounts were and are substantially true, but the fact is that the machines are more capable than what they were initially given credit for, and the percentage of cases that can be handled with automation fairly well approximates the 80/20 rule.

In this paper, we present an early stage prototype medical interpreter system that is based on lessons learned in developing successful automated coding and abstracting systems and on the core infrastructure and techniques used in these systems. Specific techniques include leveraging standards based multi-lingual medical nomenclatures and clinical ontology systems, machine awareness of difficult situations, explanatory meta-knowledge, and an interactive environment that emphasizes the strengths of both the human and machine participants and mitigates the weaknesses of each.

1 Introduction

The task of medical interpretation is demanding and difficult, and although U.S. hospitals that receive federal funds are required to provide interpreter services, the demands on the system are generally beyond the availability of qualified interpreters. Less than one fourth of U.S. hospitals have professionally trained interpreters and among these, many have no training in medical terminology. [Loviglio, 2004] After noting that well-to-do, educated patients have a relatively similar grasp of the process and content of medical care regardless of national or cultural origin, Haffner [1992] describes a variety of scenarios in which communication regarding medical treatment is far more difficult with the poor and under-educated. Karliner, Perez-Stable and Gildengorin [2004] formalize the study of medical interpreting, expand on the ad hoc observations of interpreters, and detail many of the challenges and pitfalls that befall the medical interpreter as well as the errors in medical care that may arise from inadequate cross-language and cross-cultural communications. These include hesitancy of patients to communicate fully and openly with physicians due to embarrassment or cultural norms, misunderstandings regarding offered treatments based on differing medical practices in the patient's native environment, and, in some cases, a lack of terminology by which western medical concepts can be easily translated to the patient.

As in other areas of medicine such as medical coding and abstracting, the problem of an interpreter shortage is not likely to be self-limiting. For this reason, machine translation has undeniable interest. The demands of medicine, however, require that the matter be approached in a manner different or more comprehensive than those employed in translating web pages or interpreting tourism related queries and responses. Specific needs of both physicians and patients motivate the quest for medically accurate and culturally attuned communication. Experience in building successful systems that use NLP to automate medical coding and abstracting tasks teaches that success is achieved not in trying to create a machine that replaces the human but rather is achieved by creating a machine that assists and augments the human practitioner. Specifically, the machine should offload the portions of the job that are mundane, repetitive and that can be successfully automated. In the coding and abstracting area where A-Life Medical processes the free-text transcriptions for over two million clinical encounters per month, this equates to about 70% of the total volume. [Morris, et al., 2000] A second aspect of successful human-machine collaboration is that the machine needs the ability to accurately differentiate between language-based content that it can process independently and that which requires human review and/or intervention. We call this semi-knowledge in that it corresponds to the human capability to recognize that an utterance is of importance to the task at hand even though the full intent is not comprehended. In such cases, the machine must, like a human, seek expert guidance. In this regard, the machine will address the issue to a human expert, but the strengths of the machine can be used to provide on-line help and meta-data as an aid to the human expert. This is particularly helpful in the medical field where the sheer volume of knowledge is frequently beyond the ability of humans to keep in ready memory.

In regard to the volume of knowledge, the problem in medicine is in part mitigated by the ongoing developments in the area of medical ontologies that provide for the unambiguous representation of the majority of clinical concepts. In particular, this project relies on the Systematic Nomenclature of Medicine – Clinical Terminology (SNOMED-CT®)[1] for the core, multilingual nomenclature of clinical concepts and the Clinical Document Architecture, Release 2.0 (CDA2) [Dolin, et al., 2005] for the framework by which complex clinical events and communications can be represented using the core nomenclature.

2 Motivation

More than 21 million residents of the United States speak English poorly or not at all and for more than 46 million, English is not the first language. [Karliner, Perez-Stable and Gildengorin, 2004] In many urban settings, sizable minorities of patients speak a language other than English. [Loviglio, 2004] Unless the communications needs of both the physician and patient are met, the possibility for serious medical errors is exacerbated.

2.1 What Physicians Need

All areas of physician-patient communication are important to the quality of care, but the quality of communication that is required can be divided according to those communications that are only of immediate importance during the course of the encounter and those that have durable importance beyond the temporal scope of the encounter. For example, physician directives for the patient to stand, bend, take a deep breath, etc. are in the immediate class and the accuracy of a translation (often augmented by signing, example, and physical manipulation) can be easily judged by the patient's actions. Conversely, acquiring the patient history, the review of systems, explaining diagnoses and prescribing medications and a course of treatment

[1] This material includes SNOMED Clinical Terms® (SNOMED CT®), which is used by permission of the College of American Pathologists. ©2002-2006 College of American Pathologists. All rights reserved. SNOMED CT has been created by combining SNOMED RT® and a computer based nomenclature and classification known as Clinical Terms Version 3, formerly known as the Read Codes Version 3, which was created on behalf of the U.K. Department of Health and is a Crown Copyright. SNOMED and SNOMED CT are registered trademarks of the College of American Pathologists.

have import that continues beyond the time scope of the encounter in that they become part of the permanent record, are a basis for both current and future medical decision making, and are critical to accurate completion of the course of care. Further, it is the physician's responsibility, as the care provider, to ensure that the patient has been understood and that the patient understands the nature of their condition and the planned course of treatment. Without a means to validate the communications that represent what we are calling the durable aspects, the physician can neither be sure nor give assurance that the communications have been accurate, and that the course of treatment is appropriate.

Another important area where medical interpretation services are needed is physician-to-physician communication. Telemedicine is on the rise in the United States [Bauer, 2002] and worldwide. [Sood, Bhatia, 2005] Many applications of telemedicine involve communication between physicians located in different countries. [Wachter, 2006] Effective physician-to-physician communication usually requires proficiency in nuances medical terminology and can be challenging even for physicians who are fluent in lay language. [Bruzzi, 2006]

2.2 What patients need

Although communication that is primarily physician directed with yes/no or multiple-choice patient responses can cover a lot of territory, there are several areas for which it is necessary that the patient be able to have a more comprehensive input. These include the expression of concerns about the severity and prospective outcome relative to their medical condition, and communication of issues relative to their life situation that contributed to their condition or that may affect their ability to follow medical instructions. Karliner, et al. [2004] found that even when using interpreters, physician satisfaction levels with regard to their ability to elicit exact symptoms, explain treatments, elicit treatment preferences, and empower patients with regard to their own care was far lower than for the physician's satisfaction with their ability to diagnose and treat the medical condition. During medical encounters in which the physician and patient speak the same language, the physician may likely initiate dialogue on these topics with open-ended questions such as "How did this happen?", "Do you have any other questions?", "Does this concern you?", and the like. Because the answers to these questions may be complex and may be influenced by cultural sensitivities [Hudelson, 2005], cultural context must be accounted for in the design of an automated medical interpreter system.

2.3 Statistical translation systems

Statistical translations systems have been developed for many language pairs, but due to the nature of the available parallel training corpora one language in most pairs is English. [Waibel, et al., May 2004] Further, statistical translation systems rely on large parallel corpora for training and these may not be available for applications in clinical medicine. Advances in the general state of machine translation can be tracked in the results of the NIST Machine Translation Evaluations. [NIST, 2005] [Papineni, 2002] [Zhang, Vogel and Waibel, 2004] A more complete analysis that includes measures of adequacy, fluency, and meaning maintenance can be found in Eck and Hori [2005], who provide the following medical example that is illustrative of the unevenness in these three areas of measurement and demonstrates the need for other methods beyond straight statistical translation.

Reference	**i would like to have an allergy test please**
Translation 1	i would like to have an allergy test please
Translation 2	could you check I am allergic
Translation 3	i would like to make a
Translation 4	allergic to order room service please

Statistical systems can also be time and resource intensive [Fung, et al., 2004] such that quality must be sacrificed for speed in applications that require near real-time response. [Peterson, 2006]

2.4 Interlingua translation systems

Interlingua approaches for clinical applications seem to have a preferred status, in part due to the constrained nature of clinical speech and in part due to the ability to provide a structured back-translation from the interlingua to the physician's language for confirmation of adequacy, and in part due to the fact that the interlingua provides a formally represented, deep analysis of meaning. [Schultz, et al., 2004] Two approaches to interlingua are common:

1. a formal representation of meaning [Bouillon, 2005]
2. a natural language, usually English, as interlingua. [Waibel, et al., May 2004]

With regard to using a formal interlingua, mapping speech to an unambiguous formal representation that can be validated by the speaker provides the requisite accuracy and a basis for accurate translation, but the time and expense required to build such a system for all patient languages is prohibitive. Secondly, a formal interlingua will not easily capture many nuances of natural language.

The use of a natural language as the interlingua provides the ability to represent a greater range of nuance, although all languages have subtleties that cannot easily be translated. Natural languages, however, introduce the problem of double translation errors. Further, the interlingua may be a language or format inaccessible to either speaker in a conversation, and so there is no way for either speaker to validate.

3 Approach

Our approach, currently designated as Accultran/Med or just Accultran (for Accurate, Acculturated Translator) is based, both philosophically and in terms of implementation, on LifeCode® NLP system that has been developed at A-Life Medical for coding and abstracting clinical documents. A complete description is beyond the scope of this paper but is available in [Heinze, et al. 2001]. Automated Speech Recognition (ASR) is performed using the SpeechMagic™ system from Philips, which is currently available for twenty-three languages. Non-CDA2/SNOMED-CT translations are via AltaVista Babelfish.

Many of the techniques in our approach are established in the practice. Particularly we note the use of physician directed communication with yes/no patient responses, back-translation on the physician side, and the use of multiple choice answer selections for patient responses. [Kazunori, et al., 2006] Beyond this, we are exploring the use of CDA2 and SNOMED-CT as the interlingua for use in those portions of the encounter where clinical accuracy is essential, the use of semi-knowledge for recognizing when an encounter is potentially moving in directions where cross-cultural communications problems may arise, and the use of patient waiting time for patient directed acculturation based on patient complaint information collected upon presentation. The primary emphasis here will be on the use of CDA2 and SNOMED-CT.

Based on the previous observations regarding physician and patient communication needs during a clinical encounter, we divide the clinical encounter into the following aspects: 1) establishing rapport; 2) chief complaint; 3) history; 4) review of systems; 5) physical examination; 6) diagnoses; 7) procedures; 8) medications; 9) instructions. Except for item 1, these all correspond to sections of the traditional clinical note or report and as such have extensive representations in CDA2 and SNOMED-CT. The Continuity of Care Document (CCD), a current effort to harmonize the ASTM and CDA2 standards in this realm, attempts to focus just on these elements using CDA2 representation capabilities. CDA2 is primarily declarative with some capabilities to represent contingencies. This is essentially what is needed for presenting information, but much of the encounter requires query and response.

The core of Accultran resides in the capability of the NLP engine to determine the appropriate context for each physician utterance and to appropriately process and route the content of the utterance. The overall communications flow for the system is illustrated in Figure 1, showing that upon receiving and processing an utterance from the physician, the NLP engine can choose one of several courses of action:

(1) Utterances that contain clinical questions or clinical statements for the patient to affirm or deny or instructions are: (a) converted to CDA2 and are (b) processed by a style sheet that produces the question/statement (c) first for physician validation and then (d) mapped to the patient language with, as needed, a request to affirm or deny.

(2) Utterances with content that cannot be converted to CDA2 are (a) routed to a general machine translation system, (b) optionally with back-translation and physician approval before (c) presentation to the patient.

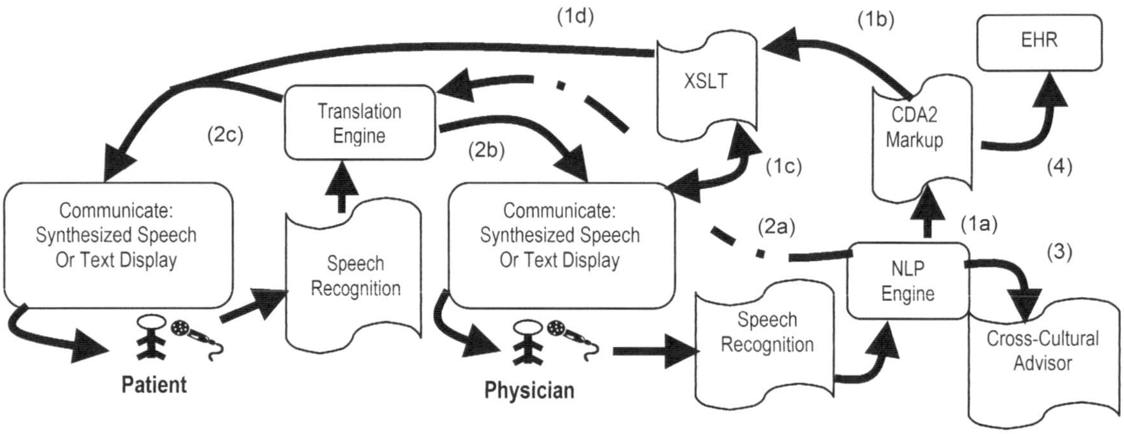

Figure 1: Framework for automated medical interpretation with cultural queues and EHR construction.

(3) Utterances that contain references to subject matter that is deemed culturally sensitive or subject to misunderstanding will trigger the Cross-Cultural Advisor.

(4) As the encounter progresses, the NLP engine appropriately directs information to the EHR via CDA2 for later physician review, and, as needed, revision.

The Cross-Cultural Advisor (CCA) module is a key feature. Technically it is based on the NLP engine's capability for recognizing and flagging clinical content that requires special attention beyond what the NLP system can independently provide. In this case, the flags are associated with warnings related to subject matter that is known to have either cultural sensitivities for patients in the target language group or that is difficult to translate into the target language. Options that the CCA could present to the physician for any particular flag would include warnings with explanation of the sensitivity, pre-formulated queries or informational presentations that are designed to mitigate any misunderstandings, or advise that a human interpreter be involved. In cases where the services of a human interpreter are called for, the CCA identified topic can be used to select, when available, an interpreter with training or skills appropriate to the case at hand. This can be particularly useful when Video Medical Interpretation (VMI) capabilities are used and there is a pool of remote interpreters from which to select.

Actual extracts (labels and annotation added) of a simplified encounter follow.

```
Physician: "Good morning."
Acccultran: "Good day."
Physician: "Okay."
Acccultran: "Buonas díaz."
Patient: "Buonas díaz."
Acccultran: "Good day."
Physician: "What seems to be
the problem?"
Acccultran: "What seems to be
the problem?"
Physician: "Okay."
Acccultran: "¿Qué se parece
ser el problema?"
Patient: "Caí en la ducha y
lastimé mi muñeca." (patient
points to left wrist)
Acccultran: "I fell in the
shower and hurt my wrist."
Physician: "You fell in the
shower and you have pain in
your left wrist?"
Acccultran: "Please confirm:
fall in shower?"
Physician: "Okay."
Acccultran: "Confirme por fa-
vor: ¿caída en ducha?"
Patient: "Si."
Acccultran: "Yes."
Acccultran: "Please confirm:
left wrist pain?"
```

Physician: "Okay."
Accultran: "Confirme por favor: ¿dolor izquierdo estructura de la región de la muñeca?"
Patient: "Si."
Accultran: "Yes."
Physician: "Do you have any other pain?"
Accultran: "Do you have other pain?"
Physician: "Okay."
Accultran: "¿Usted tiene otro dolor?"
Patient: "No."
Accultran: "No."

Examination follows a similar course of dialogue.

Physician: "You have a bruise on the left forearm and a mild sprain of the left wrist."
Accultran: "You have: left forearm contusion; mild left wrist sprain. Do you understand?"
Physician: "Okay."
Accultran: "Usted tiene: contusion izquierdo del antebrazo. ¿Usted entiende?"
Patient: "Si."
Accultran: "Yes."
Accultran: "Usted tiene: esguince izquierdo leve estructura de la región de la muñeca. ¿Usted entiende?"
Patient: "Si."
Accultran: "Yes."
Physician: "Instructions: no lifting for one week; take Tylenol 650 mg qid for two days."
Accultran: "Instructions: no lifting; duration: 1 week. Medication: Tylenol, quantity: 650 millegrams, frequency: every 6 hours, duration: 2 days."
Physician: "Okay."

```
<section>
 <title>Question</title>
 <entry>
  <observation>
   <code
     codeSystemName="SNOMED CT"
     displayName="Chief Complaint" />
    <value xsi:type="CD"
     code="44188002"
     codeSystemName="SNOMED CT"
     displayName="fall in shower">
    </value>
  </observation>
  <observation>
     codeSystemName="SNOMED CT"
     displayName="Chief Complaint" />
    <value xsi:type="CD"
     code="22253000"
     codeSystemName="SNOMED CT"
     displayName="pain">
      <qualifier>
        <name
           displayName="finding site" />
        <value
           codeSystemName="SNOMED CT"
           displayName="wrist" />
      </qualifier>
      <qualifier>
        <name
           displayName="laterality" />
        <value
           codeSystemName="SNOMED CT"
           displayName="left" />
      </qualifier>
    </value>
  </observation>
 </entry>
</section>
```

Figure 2: Abbreviated SNOMED CT for Queries

Accultran: "Instrucciones: ninguna elevación; duración: 1 semana. ¿Usted entiende?"
Patient: "Si."
Accultran: "Yes."
Accultran: "Medicación: Tylenol, cantidad: 650 millegrams, frecuencia: cada 6 horas, duración: 2 días. ¿Usted entiende?"
Patient: "Si."
Accultran: "Yes."
Physician: "Do you have any other concerns?"
Accultran: "Do you have any other concerns?"
Physician: "Okay."
Accultran: "¿Usted tiene otras preocupaciones?"
Patient: "Estoy asustado para mi bebé."(Patient points to abdomen) "Mi marido está eno-

```
jado que pude haber lastimado
al bebé."
Accultran: "I'm afraid for
may baby.  My husband is an-
gry that I may have hurt the
baby."
Physician: "Are you preg-
nant?"
Acultran: "Are you preg-
nant?"
Physician: "Okay."
Acultran: "¿Es usted em-
barazado?"
Patient: "Si."
Acultran: "Yes."
Physician: "Is your husband
angry with you?"
Acultran: "Warning: …" (Ac-
cultran produces a cultural
warning relative to the im-
portance of bearing children
in Hispanic cultures, marital
relations…. The decision is
made to involve an inter-
preter with skills in preg-
nancy and domestic issues.)
```

4 Discussion

As stated in the Introduction, Accultran is still an early prototype. The NLP engine has been evaluated at Partners Healthcare for mapping clinical free-text to CDA2. Publication of these results is expected in the near future. Development of the translation aspects is not yet sufficiently mature for field testing. Per work at A-Life, we currently see a number of strengths and several particular shortcomings with regard to CDA2 and SNOMED-CT as a framework for documenting and communicating clinical encounters. The strengths are in the coverage of medical concepts, the ability to formally assemble concepts in a coherent representation of an encounter, and the ability to easily map that formal representation to a variety of applications via XSLT (XML Style Sheets) and alternate language representation. However, although nomenclatures such as SNOMED-CT provide coverage for concepts such as embarrassment, inappropriate behavior, identification of cultural and value components related to pain management etc., they do not provide information or insights into the actual cultural components that affect these concepts. The cultural components must be developed separately and added to the system as metadata attached to specific semi-knowledge entries with attached flags and helps. This notion can be further expanded so as to use the considerable waiting time that patients typically experience in medical settings. During this time the patient would interact with the system, which would provide language and culture specific materials to educate and acculturate the patient.

Finally, and of no small import, SNOMED-CT is currently only available in two versions of English (US and UK), German and Spanish. In the US, at least, Spanish would be one of the primary languages in need. Although there are no current plans for complete and official versions of SNOMED-CT in languages other than those just noted (personal communication with author's SNOMED account manager), our requirements are for only a limited subset of the terms that could be independently translated in a commercial setting.

5 Conclusion

The difficulty of medical interpreting and the potential medical consequences should not be underestimated. Aside from the difficulties in sheer vocabulary size and multi-lingual representation, there are the added complications of diverse cultures. We have presented an architecture that addresses the issues of medical accuracy and cultural sensitivity. Although the use of such a system requires some patience and acclimation on the part of both medical practitioners and patients, the cost is small as compared to that of any morbidity or mortality that could result from inaccurate communication.

References

Bauer JC 2002. Insights on Telemedicine: How Big Is the Market? *Journal of Healthcare Information Management* 16(2): 10-11.

Bouillon P, Rayner M, Chatzichrisafis N, Hockey BA, Santaholma M, Starlander M, Isahara H, Kanzaki K, Nakao Y. May 2005. A generic Multi-Lingual Open Source Platform for Limited-Domain Medical Speech Translation. *Proceedings of the tenth Conference on European Association of Machine Translation.* Budapest, Hungary. 5-58.

Bruzzi JF 2006. The Words Count — Radiology and Medical Linguistics. *The New England Journal of Medicine* 354(7): 665-667.

Dolin RH, Alshhuler L, Boyer S, Beebe C, Behlen FM, Biron PV, Shabo A, editors. 2005. HL7 Clinical Document Architecture, Release 2.0. ANSI-approved HL7 Standard, May 2005. Ann Arbor, MI: Health Level Seven, Inc.

Eck M, Hori C. 2005. Overview of the IWSLT 2005 Evaluation Campaign. *International Workshop on Spoken Language Translation.* Pittsburgh, PA.

Flores G. 2005. The Impact of Medical Interpreter Services on the Quality of HealthCare: A Systematic Review. *Medical Care Research and Review,* 62(3): 255-299.

Fung P, Yi L, Yongsheng Y, Shen Y, Wu D. October 2004. A Grammar-Based Chinese to English Speech System for Portable Devices. *Interspeech-ICSLP.* Jeju, Korea.

Haffner L. September 1992. Translation is Not Enough: Interpreting in a Medical Setting. *Western Journal of Medicine,* 157(3).

Heinze DT, Morsch ML, Sheffer RE, Jimmink M, Jennings MA, Morris WC, Morsch AE. 2001. Life-Code: A Deployed Application for Automated Medical Coding. *AI Magazine,* 22(2): 76-88.

Hudelson P. 2005. Improving patient-provider communication: Insights from interpreters. *Family Practice,* 22(3): 311-316.

Imoto K, Sasajima M, Shimomori T, Yamanaka N, Yajima M, Masai Y. 2006. A multi modal supporting tool for multi lingual communication by inducing partner's reply. *Proceedings of the 11th International Conference on Intelligent User Interfaces.* Sydney, Australia.

Karliner LS, Perez-Stable EJ, Gildengorin G. February 2004. The Language Divide: The Importance of Training in the Use of Interpreters for Outpatient Practice. *Journal of General Internal Medicine,* 19(2): 175ff.

Loviglio J. November 22, 2004. Interpreters Lower Risks in Hospitals. Associated Press story available at http://ap.lancasteronline.com/4/hospital_babel Accessed Feb 28, 2006.

Morris WC, Heinze DT, Warner Jr. HR, Primack A, Morsch AE, Sheffer RE, Jennings MA, Morsch ML, Jimmink M. 2000. Assessing the accuracy of an automated coding system in emergency medicine. *Proceedings of the AMIA 2000 Annual Symposium.* Los Angeles, CA. 595-599.

NIST 2005 Machine Translation Evaluation Official Results. Version 3. August 1, 2005. http://www.nist.gov/speech/tests/mt/mt05eval_official_results_release_20050801_v3.html Accessed Feb 28, 2006.

Papineni K, Roukos S, Ward T, Zhu WJ. July 2002. Bleu: A Method for Automatic Evaluation of Machine Translation. *Proceedings of the 40th Annual Meeting of the Association for Computational Linguistics.* Philadelphia, PA.

Peterson R. Jan 24, 2006. IBM Strives for Superhuman Speech Tech. *Pcmag.com* http://www.pcmag.com/article2/0,1895,1915071,00.asp Accessed Feb 28, 2006.

Schultz T, Alexander D, Black AW, Peterson K, Suebvisai S, Waibel A. 2004. A Thai Translation Stystem for Medical Dialogs. *Proceedings: Human Language Technologies.* Boston, MA.

Sood SP, Bhatia JS. 2005. Development of telemedicine technology in India: "Sanjeevani"-An integrated telemedicine application. *Journal of Postgraduate Medicine,* 51(4): 308-311.

Starlander M, Bouillon P, Rayner M, Chatzichrisafis N, Hockey BA, Isahara H, Kanzaki K, Nakao Y, Santaholma M. 2005. Breaking the Language Barrier: Machine Assisted Diagnosis using the Medical Speech Translator. *Proceedings of the XIX Internation Congress of the European Federation for Medical Informatics MIE.* Geneva, Switzerland.

Wachter RM 2006. The "Dis-location" of U.S. Medicine — The Implications of Medical Outsourcing. *The New England Journal of Medicine,* 354(7): 661-665

Waibel A, Schultz T, Vogel S, Fügen C, Honal M, Kolss M, Reichert J, Stüker S. May 2004. Towards Language Portability in Statistical Speech Translation. *Special Session on Multilinguality in Speech Processing, Proceedings of the IEEE International Conference on Acoustics, Speech, and Signal Processing.* Montreal, Canada.

Zhang Y, Vogel S, Waibel A. May 2004. Interpreting BLEU/NIST Scores: How Much Improvement Do We Need to Have a Better System? *Proceedings of LREC 2004.* Lisbon, Portugal.

Language Engineering and the Pathway to Healthcare: A user-oriented view

Harold Somers
School of Informatics
University of Manchester
PO Box 88
Manchester M61 0QD, England
Harold.Somers@manchester.ac.uk

Abstract

This position paper looks critically at a number of aspects of current research into spoken language translation (SLT) in the medical domain. We first discuss the user profile for medical SLT, criticizing designs which assume that the doctor will necessarily need or want to control the technology. If patients are to be users on an equal standing, more attention must be paid to usability issues. We focus briefly on the issue of feedback in SLT systems, pointing out the difficulties of relying on text-based paraphrases. We consider the delicate issue of evaluating medical SLT systems, noting that some of the standard and much-used evaluation techniques for all aspects of the SLT chain might not be suitable for use with real users, even if they are role-playing. Finally, we discuss the idea that the "pathway to healthcare" involves much more than a face-to-face interview with a medical professional, and that different technologies including but not restricted to SLT will be appropriate along this pathway.

1 Introduction

The doctor–patient consultation is a central element of the "pathway to healthcare", and with language problems recognised as the single most significant barrier on this pathway, spoken-language translation (SLT) of doctor–patient dialogues is an obvious and timely and attractive application of language technology. As Bouillon et al. (2005) state, the task is both useful and manageable, particularly as interactions are highly constrained, and the domain can be divided into smaller domains based on symptom types. In this position paper, we wish to discuss a number of aspects of this research area, and suggest that we should broaden our horizons to look beyond the central doctor–patient consultation to consider the variety of interactions on the pathway to healthcare, and beyond the confines of SLT as an appropriate technology for patient–provider communication.

In particular we want to stress the importance of the users – both practitioners and patients – in the design, especially considering computer- and conventional literacy. We will argue that the pathway to healthcare involves a range of communicative activities requiring different language skills and implying different technologies, not restricted to SLT. We will comment on the different situations which have been targeted by research in this field so far, and the impact of different target languages on research, and how the differing avilability of resources and software influences research. We also need to consider more carefully the design of the feedback and verification elements of systems, and the need for realistic evaluations.

2 Who are the users?

We start by looking at the assumed profile of users of medical SLT systems. Systems that have been developed so far can be divided into those for use in the doctors office – notably, MedSLT (Rayner and

Bouillon, 2002), CCLINC (Lee et al., 2002), and (honourable mention) the early work done at CMU (Tomita et al., 1988)[1] – and those for use for first contact with medical professionals "in the field", developed under DARPA's CAST programme:[2] MASTOR (Zhou et al., 2004), Speechalator (Waibel et al., 2003), Transonics (Narayanan et al., 2004) and SRI's system (Precoda et al., 2004). This distinction mainly motivates differences in hardware, overall design, and coverage, but there may be other more subtle differences that result especially from the situation in which it was envisaged that the CAST systems would be used.

Some descriptions of the systems talk of "doctors" and "patients" though others do use more inclusive terms such as "medical professional". A significant common factor in the descriptions of the systems seems to be that it is the doctor who controls the device. This may be because it can only handle one-way translation, as is the case of MedSLT, "... the dialogue can be mostly initiated by the doctor, with the patient giving only non-verbal responses" (Bouillon et al., 2005), or may be an explicit design decision:

> There is, however, an assymmetry in the dialogue management in control, given the *desire* for the English-speaking doctor *to be in control* of the device and the primary "director" of the dialog. (Ettelaie et al., 2005, 89) [emphasis added]

It is understandable that as a regular user, the medical professional may *eventually* have more familiarity with the system, but this should be reflected in there being *different* user-interfaces (see Somers and Lovel 2003). We find regrettable however the assumption that "the English speaker [...] is expected to have greater technological familiarity" (Precoda et al., 2004, 9) or that

> the medical care-giver will maintain the initiative in the dialogue, will have sole access to the controls and display of the translation device, and will operate the push-to-talk controls for both him or herself and the [P]ersian patient. (Narayanan et al., 2004, 101)

In fact, although the early use of computers in doctor–patient consultations was seen as a threat, more recently the help of computers to increase communication and rapport has begun to be recognised (Mitchell and Sullivan, 2001). This may be at the expense of patient-initiated activities however, and many practitioners are suspicious of the negative impact of technology on relationships with patients, especially inasmuch as it increases the perceived power imbalance in the relationship.

Figure 1, a snapshot from Transonics demo,[3] leaves in no doubt who is in control.

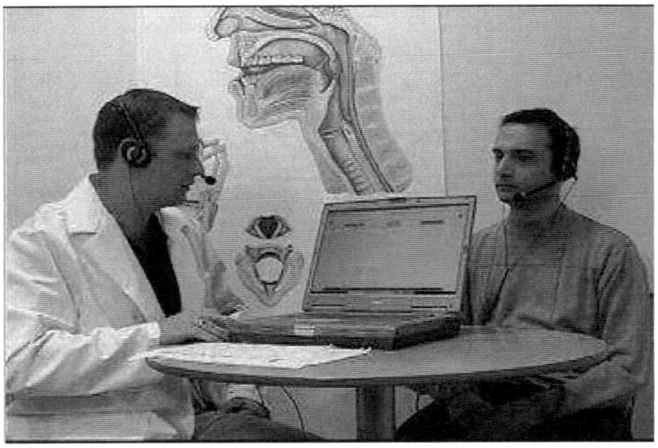

Figure 1: Snapshot from Transonics' demo movie. The patient is not even allowed to see the screen!

Equipment whose use and "ownership" can be equally shared between the participants goes some way to redressing the perceived power-balance in the consultation. We have evidence of this effect in ongoing experiments comparing (non-speech) communication aids on laptops and tablet PCs: with the laptop, controlled by a mouse or mouse-pad, the practitioner tends to take the initiative, while with the tablet, which comes with a stylus, the patient takes the lead. Bouillon et al. (2005) comment that "patients [...] will in general have had no previous exposure to speech recognition technology, and may be reluctant to try it." On the other hand, patients also have suffered from failed consultations

[1] We give here one indicative reference for each system.

[2] Formerly known as Babylon. See www.darpa.mil/ipto/programs/cast/.

[3] http://sail.usc.edu/transonics/demo/transedit02lr.mov

which break down through inability to communicate, and in our experience are pleased to be involved in experiments to find alternatives. In our view, one should not underestimate patients' adaptability, or their potential as users of technology on an equal status with the practitioners.

This being the case, we feel that some effort needs to be devoted to usability issues. We will return to this below, but note that text-based interfaces are not appropriate for users with limited literacy (which may be due to low levels of education, visual impairment, or indeed the lack of a written standard for the language). Use of images and icons also needs to be evaluated for appropriateness, an issue not addressed in any of the reports on research in medical SLT that we have read. For example, Bouillon et al. (2005) show a screenshot which includes the graphic reproduced in Figure 2. The text suggests that the user (i.e. the doctor?) can click on the picture to set the topic domain. It is not clear why a graphic is more suitable for the doctor-user than a drop-down text menu; there is no mention of whether the patient is encouraged to use the diagram, but if so one wonders for what purpose, and if it is the best choice of graphic. Research (e.g. by Costantini et al. 2002) suggests that multimodal interfaces are superior to speech-only systems, so there is some scope for exploration here.

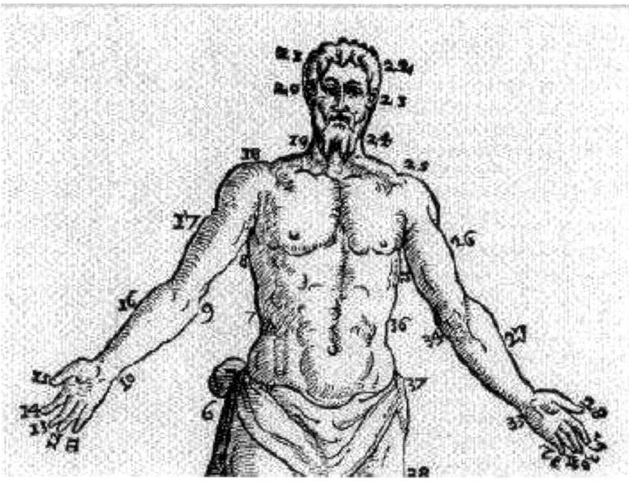

Figure 2: Graphic taken from screenshot in Bouillon et al. (2005)

Incorporating more symbolic graphics into an interface is an area of complexity, as Johnson et al. (2006) report. Iconic text-free symbols, for example to represent "please repeat", or "next question", or abstract concepts such as "very" are not always as instantly understandable as some designers think. Considering the use of symbols from AAC (augmentative and alternative communication) designed for speech-impaired disabled users by patients with limited English, we noticed that AAC symbol sets have a systematic iconicity that regular users learn, but which may be opaque to first-time (or one-time) untrained users (Johnson, 2004).

3 Feedback and verification

Translation accuracy is of course crucial in the medical domain, and sometimes problematic even with human interpreters, if not trained properly (Flores, 2005). Both speech recognition (SR) and translation are potential sources of error in the SLT chain, so it is normal and necessary to incorporate in SLT systems the provision of feedback and verification for users. The standard method for SR is textual representation, often in the form of a list of choices, for example as in Figure 3, from Precoda et al. (2004).

Figure 3: Choice of recognizer outputs, from Precoda et al. (2004:10)

For translation output, some form of paraphrase or back-translation is offered, often facilitated by the

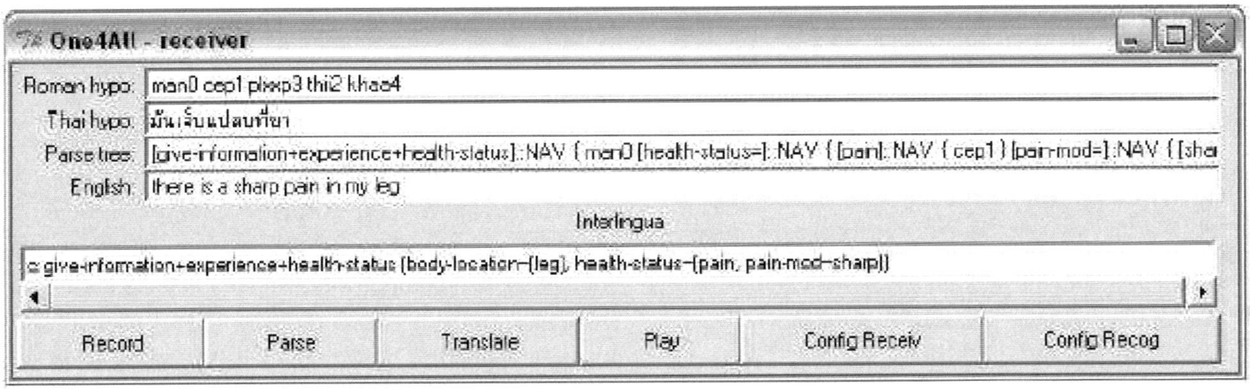

Figure 4: Choice of recognizer outputs, from Precoda et al. (2004:10)

particular design of the machine translation (MT) component (e.g. use of an interlingua representation, as in MedSLT, Speechalator). In the Transonics system, the SR accuracy is automatically assessed by the MT component: SR output that conforms to the expectations of the MT systems grammar is preferred.

For the literate English-speaking user, this approach seems reasonable, although an interface such as the one shown in Figure 4, detailing the output of the parse must be of limited utility to a doctor with no linguistics training, and we must assume that the prototype is designed more for the developers' benefit than for the end-users.

For the patient with limited or no English, the issue of feedback and verification is much more difficult. As mentioned above, and reiterated by Precoda et al. (2004), the user may not be (wholly) literate, or indeed the language (or dialect) may not have an established writing system. For some languages, displaying text in the native orthography may be an added burden. Figure 5 shows Speechalator's Arabic input screen (Waibel et al., 2003). It is acknowledged that the users must "know something about the operation of the machine", and although it is stated that the display uses the writing system of the language to be recognised, in the illustration the Arabic is shown in transcription.

Another issue concerns the ease with which a lay user can make any sense of a task in which they are asked to judge a number of paraphrases, some ungrammatical. This is an intellectual task that is difficult for someone with limited education or no experience of linguistic "games". For example, for this reason we have rejected the use of semantically unpredictable sentences (SUS) (Benoît et al., 1996) in our attempts to evaluate Somali speech synthesis (Somers et al., 2006). This leads us to a consideration of how medical SLT can best be evaluated.

4 Evaluation

MT evaluation is notoriously difficult, and SLT evaluation even more so. Most researchers agree that measures of translation fidelity in comparison with a gold-standard translation, as seen in text MT evaluation, are largely irrelevant: a task-based evaluation is more appropriate. In the case of medical SLT this presumably means simulating the typical situation that the technology will be used in, which involves patients with medical problems seeking assistance.

Since SLT is a pipeline technology, the individual components could be evaluated separately, and indeed the effects of the contributing technologies assessed (cf. Somers and Sugita 2003). Once again, literacy issues will cloud any evaluation of speech recognition accuracy that relies on its speech-to-text function, and evaluation of speech synthesis must simulate a realistic task (cf. comments on SUS, above).

Evaluations that have been reported suggest using real medical professionals and actors playing the part of patients: this scenario is well established in the medical world, where "standardized patients" (SPs) – actors trained to behave like patients – have been used since the 1960s. One problem with SPs for systems handling "low density" languages like Persian, Pashto and so on, is the need for the vol-

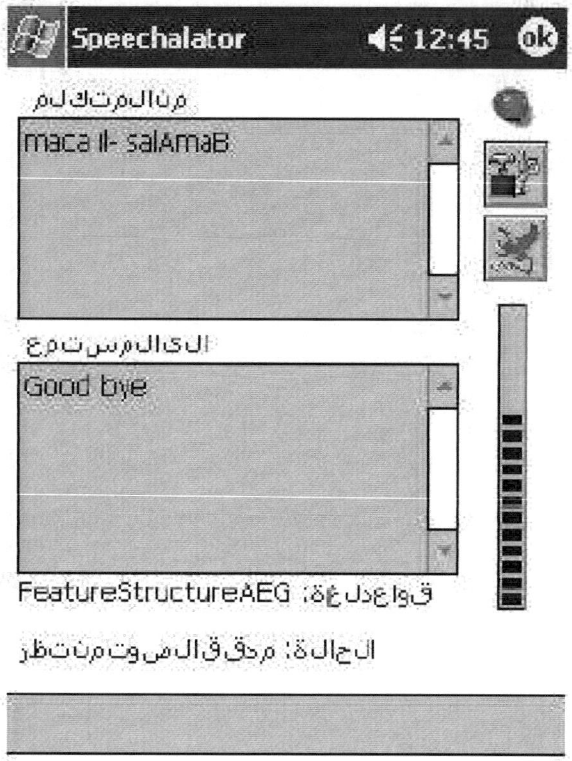

Figure 5: Speechalator's Arabic input screen (Waibel et al., 2003, 372)

unteers to understand English so that they can be trained as an SP, in conflict with the need for them to not understand English in order to give the system a realistic test. Ettelaie et al. (2005) for example report that their evaluation was somewhat compromised by the fact that two of their patient role-players did speak some English, while a third participant did not adequately understand what they were supposed to do.

Another problem is that there is no obvious baseline against which evaluations can be assessed. One could set up "with and without" trials, and measure how much and how accurately information was elicited in either mode. But this would be a waste of effort: it is widely, although anecdotally, reported that when patients with limited English arrive for a consultation where no provision for interpretation has been made, the consultations simply halt. It is also reported, as already mentioned, that human interpreters are not 100% reliable (Flores, 2005). Often, an untrained interpreter is used, whether a family member or friend that the patient has brought with them, or even another health-seeker who happens to be sitting in the waiting room. The potential for an unreliably interpreted consultation (or worse) is massive.

Ettelaie et al. (2005) mention a number of metrics that were used in their evaluation, but unfortunately do not have space for a full discussion. The principle metric is task completion, but they also mention an evaluation of a scripted dialogue, with translations evaluated against model translations using a modified version of BLEU, and SR evaluated with word-error rate. These do not seem to me to be extremely valuable evaluation techniques.

Starlander et al. (2005) report an evaluation in which the translations were judged for acceptability by native speakers. Given the goal-based nature of the task, rating for intelligibility rather than acceptability might have been more appropriate, though it is widely understood that the two features are closely related. On the positive side, Starlander et al. used only a three-point rating ("good", "ok" or "bad"): evaluations of other target languages might be subject to the problem, reported by Johnson et al. (in prep.) and by ADD REF that rating scales are highly culture-dependent, so that for example Somali participants in an evaluation of the suitability of symbols in doctor–patient communication mostly used only points 1 and 7 of a 7-point scale.

Another evaluation method[4] is to assess the number and type of translation or interpretation errors made, including whether there was any potential or actual error of clinical consequence.

As Starlander et al. (2005) say:

> In the long-term, the real question we would like to answer when evaluating the prototype is whether this system is practically useful for doctors

to which we can only add, reiterating our comments in Section 2, "... and for patients".

5 The Pathway to Healthcare

Let us move on finally to a more wide-ranging issue. "Medical SLT" is often assumed to focus on doctorpatient consultations or, as we have seen in

[4]Thanks to the anonymous reviewer for pointing this out.

the case of systems developed under the CAST programme, interactions between medical professionals and affected persons in the field. Away from that scenario, although it is natural to think of "going to the doctor" as involving chiefly an interview with a doctor, and while everything in medical practice arguably derives from this consultation, the pathway to healthcare in normal circumstances involves several other processes, all of which involve language-based encounters that present a barrier to patients with limited English. None of the medical SLT systems that have been reported in the literature address this variety of scenarios, although the website for the Phraselator (which is of course not an SLT system as such) does list a number of different scenes, such as the front desk, labour ward and so on.

In this section, we would like to survey the pathway to healthcare, and note the range of language technologies – not always speech or translation oriented – that might be appropriate at any point. The purpose of this is both to make a plea to widen our vision of what "medical SLT" covers, but also to note that SLT is not necessarily the most appropriate technology in every case.

The pathway might begin with a person suspecting that there may be something wrong with them. Many people nowadays would in this situation first try to find out something about their condition on their own, typically on the Web, though of course there is still a major "digital divide" for racial and ethnic minorities, and the poor, partly due to the langauge barriers this research is addressing. If you need this information in your own language, and you have limited literacy skills, technologies implied are multilingual information extraction. MT perhaps coupled with text simplification, with synthesized speech output. For specific conditions which may be treated at specialist clinics (our own experience is based on Somalis with respiratory difficulties) it may be possible to identify a series of frequently asked questions and set up a pre-consultation computer-mediated help-desk and interview (cf. Osman et al. 1994). See Somers and Lovel (2003) for more details.

Having decided that a visit to the doctor is indicated, the next step is to make an appointment. Appointment scheduling is the classical application of SLT, as seen in most of the early work in the field, and is a typical case of a task-oriented cooperative dialogue. Note that the "practitioner" – the receptionist in the clinic – does not necessarily have any medical expertise, nor possibly the high level of education and openness to new technology that is often assumed in the literature on medical SLT which talks of the "doctor" controlling the device.

If this is the patient's first encounter with this particular healthcare institution, there may be a process of gathering details of the patient's medivcal history and other details, done separately from the main doctor–patient consultation, to save the doctor's time. This might be a suitable application for computer-based interviewing (cf. Bachman 2003).

The next step might be the doctor–patient consultation, which has been the focus of much attention. For no doubt practical purposes, some medical SLT developers have assumed that the patients role in this can be reduced to simple responses involving yes/no responses, gestures and perhaps a limited vocabulary of simple answers at the limit. This view unfortunately ignores current clinical theory. *Patient-centred* medicine (cf. Stewart et al. 2003) is widely promoted nowadays. The session will see the doctor eliciting information in order to make a diagnosis as foreseen, but also explaining the condition and the treatment, and exploring the patients feelings about the situation. While it may be unrealistic at present to envisage fully effective support for all these aspects of the doctorpatient consultation, we feel that its purpose should be explicitly appreciated, and the limitations of current technology in this respect acknowledged.

After the initial consultation, the next step may involve a trip to the pharmacist to get some drugs or equipment. Apart from the human interaction, the drugs (or whatever) will include written instructions and information: frequency and amount of use, contraindications, warnings and so on. This is an obvious application for controlled language MT: drug dose instructions are of the same order of complexity as weather bulletins. For non-literate patients, "talking pill boxes" are already available:[5] why can't they talk in a variety of languages?

Another outcome might involve another practitioner – a nurse or a therapist – and a series of meet-

[5]Marketed by MedivoxRx. See Orlovsky (2005).

ings where the condition may be treated or managed. Apart from more scheduling, this will almost certainly involve explanations and demonstrations by the practitioner, and typically also elicitation of further information from the patient. Hospital treatment would involve interaction with a wide range of staff, again not all medical experts. If a communication device is to be used, it makes more sense for it to be under the control and "ownership" of the person who is going to be using it regularly: the patient.

6 Conclusion

Some of the comments made in this position paper may seem critical, but it has not been my intention to be negative about the field.[6] It has been my intention in this paper to draw attention to the following aspects of medical SLT which I believe so far have been somewhat neglected:

- What is the ideal user profile for medical SLT? Should the doctor control the system, or could it be seen as a shared resource?
- If the patient is also a user, devices need to be more user-friendly, taking into account cultural differences, and problems of low literacy.
- This particularly applies to feedback and verification modules in the system.
- Evaluation should focus on the ability of the technology to aid the completion of the task, from the perspective of both the practitioner and the patient.
- Evaluation methods should not involve participants in meaningless or incomprehensible tasks (such as rating nonsensical output), nor rely on skills (such as literacy) that they may lack.
- The pathway to healthcare involves more than the one-way doctor–patient dialogues covered by most systems. A wide range of technologies can be brought to bear on the problem.

[6] In particular, it should perhaps be acknowledged that in terms of practical accomplishment we have yet to match others in the field.

Acknowledgments

I would like to acknowledge the contribution to this paper of my colleagues on the CANES project, Ann Caress, Gareth Evans, Marianne Johnson, Hermione Lovel, Zeinab Mohamed. Some of the work reported here is funded by the Economic and Social Research Council (ESRC), project reference RES-000-23-0610. Thanks also to the anonymous referees for their very useful comments.

References

Bachman, J.W. 2003. 'The patient-computer interview: a neglected tool that can aid the clinician.' *Mayo Clinic Proceedings*, 78:67–78.

Benoît, Christian, Martine Grice and Valérie Hazan. 1996. 'The SUS test: A method for the assessment of text-to-speech synthesis intelligibility using Semantically Unpredictable Sentences'. *Speech Communication*, 18:381–392.

Bouillon, Pierrette, Manny Rayner, Nikos Chatzichrisafis, Beth Ann Hockey, Marianne Santaholma, Marianne Starlander, Yukie Nakao, Kyoko Kanzaki and Hitoshi Isahara. 2005. 'A generic multi-lingual open source platform for limited-domain medical speech translation'. In *Proceedings of the Tenth Conference on European Association of Machine Translation*, Budapest, Hungary, pp. CHECK

Costantini, Erica, Fabio Pianesi and Susanne Burger. 2002. 'The added value of multimodality in the NESPOLE! speech-to-speech translation system: an experimental study'. In *Fourth IEEE International Conference on Multimodal Interfaces (ICMI'02)*, Pittsburgh, PA, pp. 235–240.

Ettelaie, Emil, Sudeep Gandhe, Panayiotis Georgiou, Robert Belvin, Kevin Knight, Daniel Marcu, Shrikanth Narayanan and D. Traum. 2005. 'Transonics: A practical speech-to-speech translator for English-Farsi medical dialogues'. In *43rd Annual Meeting of the Association for Computational Linguistics: ACL-05 Interactive Poster and Demonstration Sessions*, Ann Arbor, MI, pp. 89–92.

Flores, Glenn. 2005. 'The impact of medical interpreter services on the quality of health care: a systematic review'. *Medical Care Research and Review*, 62:255–299.

Johnson, M.J. 2004. 'What can we learn from drawing parallels between people who use AAC and people whose first language is not English?' *Communication Matters*, 18(2):15–17.

Johnson, M.J., D.G. Evans and Z. Mohamed. 2006. 'A pilot study to investigate alternative communication strategies in provider-patient interaction with Somali refugees'. In *Current Perspectives in Healthcare Computing Conference*, Harrogate, England, pp. 97–106.

Johnson, M.J., G. Evans, Z. Mohamed and H. Somers (in prep.) An investigation into the perception of symbols by UK-based Somalis and English-speaking nursing students using a variety of symbol assessment techniques.

Lee, Young-Suk, Daniel J. Sinder and Clifford J. Weinstein. 2002. 'Interlingua-based English–Korean two-way speech translation of doctor–patient dialogues with CCLINC'. *Machine Translation*, 17:213–243.

Mitchell, E. and F. Sullivan. 2001. 'A descriptive feast but an evaluative famine: systematic review of published articles on primary care computing during 1980-97'. *British Medical Journal*, 322:279–282.

Narayanan, S., S. Ananthakrishnan, R. Belvin, E. Ettelaie, S. Gandhe, S. Ganjavi, P. G. Georgiou, C. M. Hein, S. Kadambe, K. Knight, D. Marcu, H. E. Neely, N. Srinivasamurthy, D. Traum, and D. Wang. 2004. 'The Transonics spoken dialogue translator: an aid for English-Persian doctor-patient interviews'. In Timothy Bickmore (ed.) *Dialogue Systems for Health Communication: Papers from the 2004 Fall Symposium*, American Association for Artificial Intelligence, Menlo Park, California, pp. 97–103.

Orlovsky, Christina. 2005. 'Talking pill bottles let medications speak for themselves'. *NurseZone.com* (online magazine), www.nursezone.com/Job/DevicesandTechnology.asp ?articleID=14396. Accessed 15 March 2006.

Osman, L., M. Abdalla, J. Beattie, S. Ross, I. Russell, J. friend, J. Legge and J. Douglas. 1994. 'Reducing hospital admissions through computer supported education for asthma patients'. *British Medical Journal*, 308:568–571.

Precoda, Kristin, Horacio Franco, Ascander Dost, Michael Frandsen, John Fry, Andreas Kathol, Colleen Richey, Susanne Riehemann, Dimitra Vergyri, Jing Zheng and Christopher Culy. 2004. 'Limited-domain speech-to-speech translation between English and Pashto'. In *HLT-NAACL 2004, Human Language Technology Conference of the North American Chapter of the Association for Computational Linguistics: Demonstrations*, pp. 9–12.

Rayner, Manny and Pierrette Bouillon. 2002. 'A flexible speech to speech phrasebook translator'. In *Proceedings of the ACL-02 Workshop on Speech-to-Speech Translation: Algorithms and Systems*, Philadelphia, PA, pp. 69–76.

Somers, Harold, Gareth Evans and Zeinab Mohamed. 2006. 'Developing speech synthesis for under-resourced languages by "faking it": an experiment with Somali'. In *Proceedings of LREC: 5th Conference on Language Resources and Evaluation*, Genoa.

Somers, Harold and Hermione Lovel. 2003. 'Computer-based support for patients with limited English'. In *Association for Computational Linguistics EACL 2003, 10th Conference of The European Chapter, Proceedings of the 7th International EAMT Workshop on MT and other language technology tools*, Budapest, pp. 41–49.

Somers, Harold and Yuriko Sugita. 2003. 'Evaluating commercial spoken language translation software.' In *MT Summit IX: Proceedings of the Ninth Machine Translation Summit*, New Orleans, pp. 370–377.

Starlander, Marianne, Pierrette Bouillon, Manny Rayner, Nikos Chatzichrisafis, Beth Ann Hockey, Hitoshi Isahara, Kyoko Kanzaki, Yukie Nakao and Marianne Santaholma. 2005. 'Breaking the language barrier: machine assisted diagnosis using the medical speech translator'. In *Proceedings of the XIX International Congress of the European Federation for Medical Informatics*, Geneva, Switzerland.

Stewart, Moira, Judith Belle Brown, W. Wayne Weston, Ian R. McWhinney, Carol L. McWilliam and Thomas R. Freeman. 2003. *Patient-Centered Medicine: Transforming the Clinical Method* (2nd ed.). Radcliffe, Abingdon, Oxon.

Tomita, Masaru, Marion Kee, Hiroaki Saito, Teruko Mitamura and Hideto Tomabechi. 1988. 'The universal parser compiler and its application to a speech translation system'. In *Second International Conference on Theoretical and Methodological Issues in Machine Translation of Natural Languages*, Pittsburgh, Pennsylvania, pages not numbered.

Waibel, Alex, Ahmed Badran, Alan W. Black, Robert Frederking, Donna Gates, Alon Lavie, Lori Levin, Kevin Lenzo, Laura Mayfield Tomokiyo, Jürgen Reichert, Tanja Schultz, Dorcas Wallace, Monika Woszczyna, and Jing Zhang. 2003. 'Speechalator: two-way speech-to-speech translation on a consumer PDA'. In *Proceedings of EUROSPEECH 2003, 8th European Conference on Speech Communication and Technology*, Geneva, pp. 369–372.

Zhou, Bowen, Daniel Déchelotte and Yuqing Gao. 2004. 'Two-way speech-to-speech translation on handheld devices'. In *INTERSPEECH 2004 – ICSLP, 8th International Conference on Spoken Language Processing*, Jeju Island, Korea, pp. 1637–1640.

Converser™: Highly Interactive Speech-to-Speech Translation for Healthcare

Mike Dillinger
Spoken Translation, Inc.
Berkeley, CA, USA 94705
mike.dillinger@spokentranslation.com

Mark Seligman
Spoken Translation, Inc.
Berkeley, CA, USA 94705
mark.seligman@spokentranslation.com

Abstract

We describe a highly interactive system for bidirectional, broad-coverage spoken language communication in the healthcare area. The paper briefly reviews the system's interactive foundations, and then goes on to discuss in greater depth our Translation Shortcuts facility, which minimizes the need for interactive verification of sentences after they have been vetted. This facility also considerably speeds throughput while maintaining accuracy, and allows use by minimally literate patients for whom any mode of text entry might be difficult.

1 Introduction

Spoken Translation, Inc. (STI) of Berkeley, CA has developed a commercial system for interactive speech-to-speech machine translation designed for both high accuracy and broad linguistic and topical coverage. Planned use is in situations requiring both of these features, for example in helping Spanish-speaking patients to communicate with English-speaking doctors, nurses, and other healthcare staff.

The twin goals of accuracy and broad coverage have until now been in opposition: speech translation systems have gained tolerable accuracy only by sharply restricting both the range of topics which can be discussed and the sets of vocabulary and structures which can be used to discuss them. The essential problem is that both speech recognition and translation technologies are still quite error-prone. While the error rates may be tolerable when each technology is used separately, the errors combine and even compound when they are used together. The resulting translation output is generally below the threshold of usability – unless restriction to a very narrow domain supplies sufficient constraints to significantly lower the error rates of both components.

STI's approach has been to concentrate on interactive monitoring and correction of both technologies.

First, users can monitor and correct the speaker-dependent speech recognition system to ensure that the text, which will be passed to the machine translation component, is completely correct. Voice commands (e.g. **Scratch That** or **Correct <incorrect text>**) can be used to repair speech recognition errors. While these commands are similar in appearance to those of IBM's ViaVoice or ScanSoft's Dragon NaturallySpeaking dictation systems, they are unique in that they will remain usable even when speech recognition operates at a server. Thus, they will provide for the first time the capability to interactively confirm or correct wide-ranging text, which is dictated from anywhere.

Next, during the MT stage, users can monitor, and if necessary correct, one especially important aspect of the translation – lexical disambiguation.

STI's approach to lexical disambiguation is twofold: first, we supply a specially controlled *back translation*, or translation of the translation. Using this paraphrase of the initial input, even a monolingual user can make an initial judgment concerning the quality of the preliminary machine translation output. To make this technique effective, we use proprietary facilities to ensure that the lexical senses used during back translation are appropriate.

In addition, in case uncertainty remains about the correctness of a given word sense, we supply a proprietary set of Meaning Cues™ – synonyms, definitions, etc. – which have been drawn from various resources, collated in a unique database (called SELECT™), and aligned with the respective lexica of the relevant machine translation systems. With these cues as guides, the user can select the preferred meaning from among those available. Automatic updates of translation and back translation then follow.

The result is an utterance, which has been monitored and perhaps repaired by the user at two levels – those of speech recognition and translation. By employing these interactive techniques while integrating state-of-the-art dictation and machine translation programs – we work with Dragon Naturally Speaking for speech recognition; with Word Magic MT (for the current Spanish system); and with ScanSoft for text-to-speech – we have been able to build the first commercial-grade speech-to-speech translation system which can achieve broad coverage without sacrificing accuracy.

2 Translation Shortcuts

In order to accumulate translations that have been verified by hand and to simplify interaction with the system, we have developed additional functionality called Translation Shortcuts™.

Shortcuts are designed to provide two main advantages:

First, re-verification of a given utterance is unnecessary. That is, once the translation of an utterance has been verified interactively, it can be saved for later reuse, simply by activating a **Save as Shortcut** button on the translation verification screen. The button gives access to a dialogue in which a convenient *Shortcut Category* for the Shortcut can be selected or created. At reuse time, no further verification will be required. (In addition to such dynamically created *Personal* Shortcuts, any number of prepackaged *Shared* Shortcuts can be included in the system.)

Second, access to stored Shortcuts is very quick, with little or no need for text entry. Several facilities contribute to meeting this design criterion.

- A *Shortcut Search* facility can retrieve a set of relevant Shortcuts given only keywords or the first few characters or words of a string. The desired Shortcut can then be executed with a single gesture (mouse click or stylus tap) or voice command.

NOTE: If no Shortcut is found, the system automatically gives access to the full power of broad-coverage, interactive speech translation. Thus, a seamless transition is provided between Shortcuts and full translation.

- A *Translation Shortcuts Browser* is provided, so that users can find needed Shortcuts by traversing a tree of Shortcut categories. Using this interface, users can execute Shortcuts even if their ability to input text is quite limited, e.g. by tapping or clicking alone.

The demonstration will show the Shortcut Search and Shortcuts Browser facilities in use. Points to notice:

- The Translation Shortcuts Panel contains the Translation Shortcuts Browser, split into two main areas, Shortcuts Categories (above) and Shortcuts List (below).
- The Categories section of the Panel shows the current selected category, for example **Conversation**, which contains everyday expressions. This category has a **Staff** subcategory, containing expressions most likely to be used by healthcare staff members. There is also a **Patients** subcategory, used for patient responses. Such categories as **Administrative topics** and **Patient's Current Condition** are also available; and new ones can be freely created.
- Below the Categories section is the Shortcuts List section, containing a scrollable list of alphabetized Shortcuts. (Various other sorting criteria will be available in the future, e.g. sorting by frequency of use, recency, etc.)
- Double clicking on any visible Shortcut in the List will execute it. Clicking once will select and highlight a Shortcut. Typing **Enter** will execute the currently highlighted Shortcut, if any.
- It is possible to automatically relate options for a patient's response to the previous staff member's utterance, e.g. by automatically going to the sibling **Patient** subcategory if the prompt was given from the **Staff** subcategory.

Because the Shortcuts Browser can be used without text entry, simply by pointing and clicking, it enables responses by minimally literate users. In the future, we plan to enable use even by completely illiterate users, through two devices: we will enable automatic pronunciation of Shortcuts and categories in the Shortcuts Browser via text-to-speech, so that these elements can in effect be read aloud to illiterate users; and we will augment Shared Shortcuts with pictorial symbols, as clues to their meaning.

A final point concerning the Shortcuts Browser: it can be operated entirely by voice commands, although this mode is more likely to be useful to staff members than to patients.

We turn our attention now to the Input Window, which does double duty for Shortcut Search and arbitrary text entry for full translation. We will consider the search facility first.

- Shortcuts Search begins automatically as soon as text is entered by any means – voice, handwriting, touch screen, or standard keyboard – into the Input Window.
- The **Shortcuts Drop-down Menu** appears just below the Input Window, as soon as there are results to be shown. The user can enter a few words at a time, and the drop-down menu will perform keyword-based searches and present the changing results dynamically.
- The results are sorted alphabetically. Various other sorting possibilities may be useful: by frequency of use, proportion of matched words, etc.
- The highest priority Shortcut according to the specified sorting procedure can be highlighted for instant execution.
- Highlighting in the drop-down menu is synchronized with that of the Shortcuts list in the Shortcuts Panel.
- Arrow keys or voice commands can be used to navigate the drop-down menu.
- If the user goes on to enter the exact text of any Shortcut, e.g. "Good morning," a message will show that this is in fact a Shortcut, so that verification will not be necessary. However, final text not matching a Shortcut, e.g. "Good job," will be passed to the routines for full translation with verification.

3 Future developments

We have already mentioned plans to augment the Translation Shortcuts facility with text-to-speech and iconic pictures, thus moving closer to a system suitable for communication with completely illiterate or incapacitated patients.

Additional future directions follow.

- **Server-based architectures:** We plan to move toward completely or partially server-based arrangements, in which only a very thin client software application – for example, a web interface – will run on the client device. Such architectures will permit delivery of our system on smart phones in the Blackberry or Treo class. Delivery on handhelds will considerably diminish the issues of physical awkwardness discussed above, and anytime/anywhere/any-device access to the system will considerably enlarge its range of uses.
- **Pooling Translation Shortcuts:** As explained above, the current system now supports both Personal (do-it-yourself) and Shared (prepackaged) Translation Shortcuts. As yet, however, there are no facilities to facilitate pooling of Personal Shortcuts among users, e.g. those in a working group. In the future, we will add facilities for exporting and importing shortcuts.
- **Translation memory:** Translation Shortcuts can be seen as a variant of Translation Memory, a facility that remembers past successful translations so as to circumvent error-prone reprocessing. However, at present, we save Shortcuts only when explicitly ordered. If all other successful translations were saved, there would soon be far too many to navigate effectively in the Translation Shortcuts Browser. In the future, however, we could in fact record these translations in the background, so that there would be no need to re-verify new input that matched against them. Messages would advise the user that verification was being bypassed in case of a match.
- **Additional languages:** The full SLT system described here is presently operational only for bidirectional translation between English and Spanish. We expect to expand the system to Mandarin Chinese next. Limited working prototypes now exist for Japanese and German, though we expect these languages to be most useful in application fields other than healthcare.

4 Conclusion

We have described a highly interactive system for bidirectional, broad-coverage spoken language communication in the healthcare area. The paper has briefly reviewed the system's interactive foundations, and then gone on to discuss in greater depth issues of practical usability.

We have presented our Translation Shortcuts facility, which minimizes the need for interactive verification of sentences after they have been vetted once, considerably speeds throughput while maintaining accuracy, and allows use by minimally literate patients for whom any mode of text entry might be difficult.

We have also discussed facilities for multimodal input, in which handwriting, touch screen, and keyboard interfaces are offered as alternatives to speech input when appropriate. In order to deal with issues related to physical awkwardness, we have briefly mentioned facilities for hands-free or eyes-free operation of the system.

Finally, we have pointed toward several directions for future improvement of the system.

MedSLT: A Limited-Domain Unidirectional Grammar-Based Medical Speech Translator

Manny Rayner, Pierrette Bouillon, Nikos Chatzichrisafis, Marianne Santaholma, Marianne Starlander
University of Geneva, TIM/ISSCO, 40 bvd du Pont-d'Arve, CH-1211 Geneva 4, Switzerland
Emmanuel.Rayner@issco.unige.ch
Pierrette.Bouillon@issco.unige.ch, Nikos.Chatzichrisafis@vozZup.com
Marianne.Santaholma@eti.unige.ch, Marianne.Starlander@eti.unige.ch

Beth Ann Hockey
UCSC/NASA Ames Research Center, Moffet Field, CA 94035
bahockey@email.arc.nasa.gov

Yukie Nakao, Hitoshi Isahara, Kyoko Kanzaki
National Institute of Information and Communications Technology
3-5 Hikaridai, Seika-cho, Soraku-gun, Kyoto, Japan 619-0289
yukie-n@khn.nict.go.jp, {isahara, kanzaki}@nict.go.jp

Abstract

MedSLT is a unidirectional medical speech translation system intended for use in doctor-patient diagnosis dialogues, which provides coverage of several different language pairs and subdomains. Vocabulary ranges from about 350 to 1000 surface words, depending on the language and subdomain. We will demo both the system itself and the development environment, which uses a combination of rule-based and data-driven methods to construct efficient recognisers, generators and transfer rule sets from small corpora.

1 Overview

The mainstream in speech translation work is for the moment statistical, but rule-based systems are still a very respectable alternative. In particular, nearly all systems which have actually been deployed are rule-based. Prominent examples are (Phraselator, 2006; S-MINDS, 2006; MedBridge, 2006).

MedSLT (MedSLT, 2005; Bouillon et al., 2005) is a unidirectional medical speech translation system for use in doctor-patient diagnosis dialogues, which covers several different language pairs and subdomains. Recognition is performed using grammar-based language models, and translation uses a rule-based interlingual framework. The system, including the development environment, is built on top of Regulus (Regulus, 2006), an Open Source platform for developing grammar-based speech applications, which in turn sits on top of the Nuance Toolkit.

The demo will show how MedSLT can be used to carry out non-trivial diagnostic dialogues. In particular, we will demonstrate how an integrated intelligent help system counteracts the brittleness inherent in rule-based processing, and rapidly leads new users towards the supported system coverage. We will also demo the development environment, and show how grammars and sets of transfer rules can be efficiently constructed from small corpora of a few hundred to a thousand examples.

2 The MedSLT system

The MedSLT demonstrator has already been extensively described elsewhere (Bouillon et al., 2005; Rayner et al., 2005a), so this section will only present a brief summary. The main components are a set of speech recognisers for the source languages, a set of generators for the target languages, a translation engine, sets of rules for translating to and from interlingua, a simple discourse engine for dealing with context-dependent translation, and a top-level which manages the information flow between the other modules and the user.

MedSLT also includes an intelligent help module, which adds robustness to the system and guides the user towards the supported coverage. The help module uses a backup recogniser, equipped with a statistical language model, and matches the results from this second recogniser against a corpus of utterances which are within system coverage and translate correctly. In previous studies, we showed that the grammar-based recogniser performs much better than the statistical one on in-coverage utterances, but worse on out-of-coverage ones. Having the help system available approximately doubled the speed at which subjects learned, measured as the average difference in semantic error rate between the results for their first quarter-session and their last quarter-session (Rayner et al., 2005a). It is also possible to recover from recognition errors by selecting a displayed help sentence; this typically increases the number of acceptably processed utterances by about 10% (Starlander et al., 2005).

We will demo several versions of the system, using different source languages, target languages and subdomains. Coverage is based on standard examination questions obtained from doctors, and consists mainly of yes/no questions, though there is also support for WH-questions and elliptical utterances. Table 1 gives examples of the coverage in the English-input headache version, and Table 2 summarises recognition performance in this domain for the three main input languages. Differences in the sizes of the recognition vocabularies are primarily due to differences in use of inflection. Japanese, with little inflectional morphology, has the smallest vocabulary; French, which inflects most parts of speech, has the largest.

Where?
"do you experience the pain in your jaw" "does the pain spread to the shoulder"
When?
"have you had the pain for more than a month" "do the headaches ever occur in the morning"
How long?
"does the pain typically last a few minutes" "does the pain ever last more than two hours"
How often?
"do you get headaches several times a week" "are the headaches occurring more often"
How?
"is it a stabbing pain" "is the pain usually severe"
Associated symptoms?
"do you vomit when you get the headaches" "is the pain accompanied by blurred vision"
Why?
"does bright light make the pain worse" "do you get headaches when you eat cheese"
What helps?
"does sleep make the pain better" "does massage help"
Background?
"do you have a history of sinus disease" "have you had an e c g"

Table 1: Examples of English MedSLT coverage

3 The development environment

Although the MedSLT system is rule-based, we would, for the usual reasons, prefer to acquire these rules from corpora using some well-defined method. There is, however, little or no material available for most medical speech translation domains, including ours. As noted in (Probst and Levin, 2002), scarcity of data generally implies use of some strategy to obtain a carefully structured training corpus. If the corpus is not organised in this way, conflicts between alternate learned rules occur, and it is hard to induce a stable set of rules. As Probst and Levin suggest, one obvious way to attack the problem is to implement a (formal or informal) elicitation strategy, which biases the informant towards translations which are consistent with the existing ones. This is the approach we have adopted in MedSLT.

The Regulus platform, on which MedSLT is based, supports rapid construction of complex grammar-based language models; it uses an example-based method driven by small corpora of disambiguated parsed examples (Rayner et al., 2003; Rayner et al., 2006), which extracts most of the structure of the model from a general linguistically motivated resource grammar. The result is a specialised version of the general grammar, tailored to the example corpus, which can then be compiled into an efficient recogniser or into a genera-

Language	Vocab	WER	SemER
English	441	6%	18%
French	1025	8%	10%
Japanese	347	4%	4%

Table 2: Recognition performance for English, French and Japanese headache-domain recognisers. "Vocab" = number of surface words in source language recogniser vocabulary; "WER" = Word Error Rate for source language recogniser, on in-coverage material; "SemER" = semantic error rate for source language recogniser, on in-coverage material.

tion module. Regulus-based recognisers and generators are easy to maintain, and grammar structure is shared automatically across different subdomains. Resource grammars are available for several languages, including English, Japanese, French and Spanish.

Nuance recognisers derived from the resource grammars produce both a recognition string and a semantic representation. This representation consists of a list of key/value pairs, optionally including one level of nesting; the format of interlingua and target language representations is similar. The formalism is sufficiently expressive that a reasonable range of temporal and causal constructions can be represented (Rayner et al., 2005b). A typical example is shown in Figure 1. A translation rule maps a list of key/value pairs to a list of key/value pairs, optionally specifying conditions requiring that other key/value pairs either be present or absent in the source representation.

When developing new coverage for a given language pair, the developer has two main tasks. First, they need to add new training examples to the corpora used to derive the specialised grammars used for the source and target languages; second, they must add translation rules to handle the new key/value pairs. The simple structure of the MedSLT representations makes it easy to support semi-automatic acquisition of both of these types of information. The basic principle is to attempt to find the minimal set of new rules that can be added to the existing set, in order to cover the new corpus example; this is done through a short elicitation dialogue with the developer. We illustrate this with a simple example.

Suppose we are developing coverage for the English → Spanish version of the system, and that the English corpus sentence "does the pain occur at night" fails to translate. The acquisition tool first notes that processing fails when converting from interlingua to Spanish. The interlingua representation is

```
[[utterance_type,ynq],
 [pronoun,you],
 [state,have_symptom],
 [symptom,pain],[tense,present],
 [prep,in_time],[time,night]]
```

Applying Interlingua → Spanish rules, the result is

```
[[utterance_type,ynq],
 [pronoun,usted],
 [state,tener],[symptom,dolor],
 [tense,present],
 [prep,por_temporal],
 failed:[time,night]]
```

where the tag `failed` indicates that the element `[time,night]` could not be processed. The tool matches the incomplete transferred representation against a set of correctly translated examples, and shows the developer the English and Spanish strings for the three most similar ones, here

```
does it appear in the morning
-> tiene el dolor por la mañana

does the pain appear in the morning
-> tiene el dolor por la mañana

does the pain come in the morning
-> tiene el dolor por la mañana
```

This suggests that a translation for "does the pain occur at night" consistent with the existing rules would be "tiene el dolor por la noche". The developer gives this example to the system, which parses it using both the general Spanish resource grammar and the specialised grammar used for generation in the headache domain. The specialised grammar fails to produce an analysis, while the resource grammar produces two analyses,

```
[[utterance_type,ynq],
 [pronoun,usted],
 [state,tener],[symptom,dolor],
```

```
[[utterance_type,ynq],[pronoun,you],[state,have_symptom],
 [tense,present],[symptom,headache],[sc,when],
 [[clause,[[utterance_type,dcl],[pronoun,you],
  [action,drink],[tense,present],[cause,coffee]]]]]
```

Figure 1: Representation of "do you get headaches when you drink coffee"

```
 [tense,present],
 [prep,por_temporal],
 [temporal,noche]]
```

and

```
[[utterance_type,dcl],
 [pronoun,usted],
 [state,tener],[symptom,dolor],
 [tense,present],
 [prep,por_temporal],
 [temporal,noche]]
```

The first of these corresponds to the YN-question reading of the sentence ("do you have the pain at night"), while the second is the declarative reading ("you have the pain at night"). Since the first (YN-question) reading matches the Interlingua representation better, the acquisition tool assumes that it is the intended one. It can now suggest two pieces of information to extend the system's coverage.

First, it adds the YN-question reading of "tiene el dolor por la noche" to the corpus used to train the specialised generation grammar. The piece of information acquired from this example is that `[temporal,noche]` should be realised in this domain as "la noche". Second, it compares the correct Spanish representation with the incomplete one produced by the current set of rules, and induces a new Interlingua to Spanish translation rule. This will be of the form

```
[time,night] -> [temporal,noche]
```

In the demo, we will show how the development environment makes it possible to quickly add new coverage to the system, while also checking that old coverage is not broken.

References

P. Bouillon, M. Rayner, N. Chatzichrisafis, B.A. Hockey, M. Santaholma, M. Starlander, Y. Nakao, K. Kanzaki, and H. Isahara. 2005. A generic multi-lingual open source platform for limited-domain medical speech translation. In *In Proceedings of the 10th Conference of the European Association for Machine Translation (EAMT)*, Budapest, Hungary.

MedBridge, 2006. http://www.medtablet.com/index.html. As of 15 March 2006.

MedSLT, 2005. http://sourceforge.net/projects/medslt/. As of 15 March 2005.

Phraselator, 2006. http://www.phraselator.com. As of 15 March 2006.

K. Probst and L. Levin. 2002. Challenges in automatic elicitation of a controlled bilingual corpus. In *Proceedings of the 9th International Conference on Theoretical and Methodological Issues in Machine Translation*.

M. Rayner, B.A. Hockey, and J. Dowding. 2003. An open source environment for compiling typed unification grammars into speech recognisers. In *Proceedings of the 10th EACL (demo track)*, Budapest, Hungary.

M. Rayner, P. Bouillon, N. Chatzichrisafis, B.A. Hockey, M. Santaholma, M. Starlander, H. Isahara, K. Kankazi, and Y. Nakao. 2005a. A methodology for comparing grammar-based and robust approaches to speech understanding. In *Proceedings of the 9th International Conference on Spoken Language Processing (ICSLP)*, Lisboa, Portugal.

M. Rayner, P. Bouillon, M. Santaholma, and Y. Nakao. 2005b. Representational and architectural issues in a limited-domain medical speech translator. In *Proceedings of TALN/RECITAL*, Dourdan, France.

M. Rayner, B.A. Hockey, and P. Bouillon. 2006. *Putting Linguistics into Speech Recognition: The Regulus Grammar Compiler*. CSLI Press, Chicago.

Regulus, 2006. http://sourceforge.net/projects/regulus/. As of 15 March 2006.

S-MINDS, 2006. http://www.sehda.com. As of 15 March 2006.

M. Starlander, P. Bouillon, N. Chatzichrisafis, M. Santaholma, M. Rayner, B.A. Hockey, H. Isahara, K. Kanzaki, and Y. Nakao. 2005. Practicing controlled language through a help system integrated into the medical speech translation system (MedSLT). In *Proceedings of the MT Summit X*, Phuket, Thailand.

S-MINDS 2-Way Speech-to-Speech Translation System

Farzad Ehsani, Jim Kimzey, Demitrios Master, Karen Sudre, David Domingo
Engineering Department
Sehda, Inc.
Mountain View, CA 94043
{farzad, jkimzey, dlm, karen, ddomingo}@sehda.com

Hunil Park
Independent Consultant
Seoul, Korea
phunil@hotmail.com

Abstract

Sehda's 2-way speech translation system, S-MINDS, interprets between provider and patient in routine medical interactions with very high accuracy. Optimizing the system for new tasks or languages requires very little data. New developments include a hybrid translation approach that allows participants to say complex or out-of-domain utterances, the expansion of hands-free functionality, and the ability to deliver the most urgent expressions instantaneously.

1 Introduction

Speech translation technology has the potential to give nurses and other clinicians immediate access to consistent, easy-to-use, and accurate medical interpretation for routine patient encounters. This could improve safety and quality of care for patients who speak a different language from that of the healthcare provider.

The most common hospital interactions are interview-style dialogs where the provider's and patient's utterances are simple and relatively predictable. Sehda's speech translation system, S-MINDS, focuses on translating in such situations with extremely high accuracy.

One key difference between S-MINDS and other speech translation systems is the amount of data required in development. Most other systems depend on a moderate amount of domain-specific data being available. If the data is not already available, it is extremely time- and labor-intensive for a developer to collect enough realistic data to effectively model a pure SMT system – even if the developer has direct access to a group of actual users for whom its system is being optimized.

For this and other reasons, Sehda focuses on rapid building and deployment of speech translation systems for tasks or languages where little or no data is available.

2 System Description

This section describes the speech recognition, translation, speech generation, interface and hardware components that make up S-MINDS.

2.1 Speech Recognition

S-MINDS uses a number of voice-independent automated speech recognition (ASR) engines, with the usage dependent on the languages and the particular domain. These engines include Nuance 8.5[i], SRI EduSpeak 2.0[ii], and Entropic's HTK-based engine.[iii]

Sehda's (internal) dialog/translation creation tools allow developers to compile and run new dialogs with any ASR engine so they do not have to be encumbered by the nuances of any particular engine.

2.2 Translation

S-MINDS processes ASR output using a combination of grammars and language models that is selected based on the task and the availability of training data.

S-MINDS first employs a semantic parser to extract the essential words and phrases from the ASR output. This information is then fed into

Sehda's proprietary interpretation engine, which matches the information against a finite set of concepts in the specified domain. The resulting translation is extremely accurate – often more accurate than the ASR output itself. However, as the name suggests, this engine does not directly translate users' utterances but interprets what they say and paraphrases their statements.

2.3 *Speech Generation*

S-MINDS uses its own voice generation system, which splices human recordings, to output most translations. If recordings do not exist for a word or phrase, S-MINDS generates the speech using a text-to-speech (TTS) engine.

S-MINDS includes a set of tools by which users can modify and augment the existing system with additional words and phrases in the field in a matter of a few minutes.

2.4 *Interface*

A variety of interface features make S-MINDS particularly easy to use in a hospital environment.

Most S-MINDS functions can be performed hands-free and eyes-free via a voice user interface (VUI) so the provider can focus on the patient and the operation of hospital equipment.

A picture viewer allows digital images to be displayed to aid communication with the patient and add clarity to the log.

Verbal or on-screen verification can be employed (with adjustable upper and lower thresholds) to put an additional check on recognition accuracy.

2.5 *Hardware*

A complete S-MINDS system contains three main hardware components: a Windows XP computer with S-MINDS software installed; a headset microphone, which the healthcare provider uses to control S-MINDS and communicate with the patient; and a telephone handset, which the patient uses to communicate with the provider.

3 Current Developments

Under a contract with DARPA, Sehda has developed a more interactive system using a combination of SMT and interpretation engines. This allows the users to speak more freely. If an utterance is too complex or too far 'out of domain' to be handled by the interpretation engine, S-MINDS falls back to the SMT engine, which returns a fairly reliable word-for-word translation of the ASR output.

The VUI in S-MINDS is being enhanced to include nearly all system control functions, reducing the need to change settings manually. In addition, users will be able to deliver some urgent expressions (such as "Hold still" or "You can breathe") instantaneously without saying a 'hotword' first.

---

[i] http://www.nuance.com/nuancerecognition/
[ii] http://www.speechatsri.com/products/eduspeak.shtml
[iii] http://htk.eng.cam.ac.uk/

Accultran: Automated Interpretation of Clinical Encounters with Cultural Cues and Electronic Health Record Generation

Daniel T. Heinze, PhD
A-Life Medical, Inc.
6055 Lusk Blvd. – Suite 200
San Diego, CA 92130
dheinze@alifemedical.com

Alexander Turchin, MD, MS
Brigham and Women's Hospital
221 Longwood Ave.
Boston, MA 02115
aturchin@partners.org

V. Jagannathan, PhD
West Virginia University and MedQuist, Inc.
235 High Street, Suite 213
Morgantown, WV 26505
juggy@medquist.com

1 Approach

Accultran/Med or just Accultran (for Accurate, Acculturated Translator) is based, both philosophically and in terms of implementation, on Life-Code® NLP system that has been developed at A-Life Medical for coding and abstracting clinical documents. Automated Speech Recognition (ASR) is performed using the SpeechMagic™ system from Philips, which is currently available for twenty-three languages.

Many of the techniques employed in Accultran are established in the practice. Particularly we note the use of physician directed communication with yes/no patient responses, back-translation on the physician side, and the use of multiple choice answer selections for patient responses. Beyond this, we are exploring the use of CDA2 and SNOMED-CT as the interlingua for use in those portions of the encounter where clinical accuracy is essential, the use of semi-knowledge for recognizing when an encounter is potentially moving in directions where cross-cultural communications problems may arise, and the use of patient waiting time for patient directed acculturation based on patient complaint information collected upon presentation. The primary emphasis here will be on the use of CDA2 and SNOMED-CT.

Based on observations regarding physician and patient communication needs during a clinical encounter, we divide the clinical encounter into the following aspects: 1) establishing rapport; 2) chief complaint; 3) history; 4) review of systems; 5) physical examination; 6) diagnoses; 7) procedures; 8) medications; 9) instructions. Except for item 1, these all correspond to sections of the traditional clinical note or report and as such have extensive representations in CDA2 and SNOMED-CT. The Continuity of Care Document (CCD), a current effort to harmonize the ASTM and CDA2 standards in this realm, attempts to focus just on these elements using CDA2 representation capabilities. CDA2 is primarily declarative with some capabilities to represent contingencies. This is essentially what is needed for presenting information, but much of the encounter requires query and response.

The core of Accultran resides in the capability of the NLP engine to determine the appropriate context for each physician utterance and to appropriately process and route the content of the utterance. The overall communications flow for the system is illustrated in Figure 1, showing that upon receiving and processing an utterance from the physician, the NLP engine can choose one of several courses of action:

(1) Utterances that contain clinical questions or clinical statements for the patient to affirm or deny or instructions are: (a) converted to CDA2 and are (b) processed by a style sheet that produces the question/statement (c) first for physician validation and then (d) mapped to the patient language with, as needed, a request to affirm or deny.

(2) Utterances with content that cannot be converted to CDA2 are (a) routed to a general machine translation system, (b) optionally with back-translation and physician approval before (c) presentation to the patient.

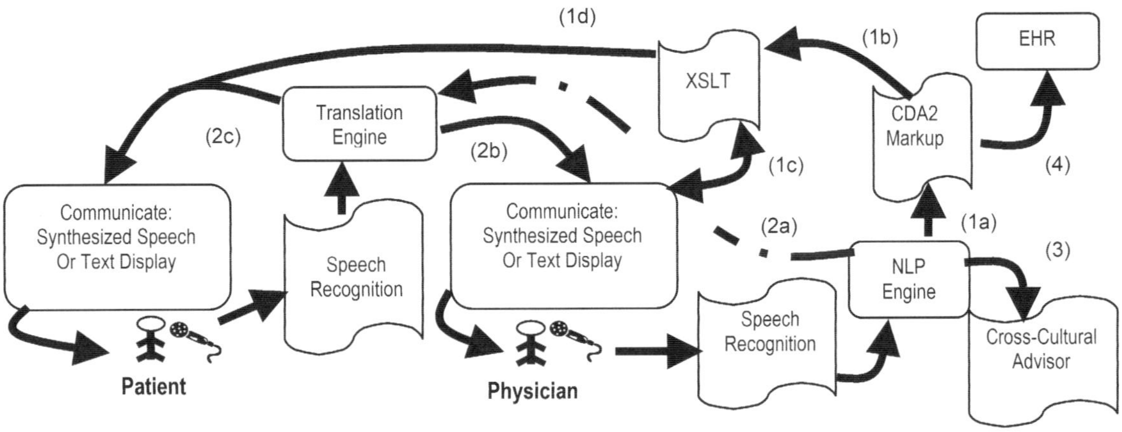

Figure 1: Framework for automated medical interpretation with cultural queues and EHR construction.

(3) Utterances that contain references to subject matter that is deemed culturally sensitive or subject to misunderstanding will trigger the Cross-Cultural Advisor.

(4) As the encounter progresses, the NLP engine appropriately directs information to the EHR via CDA2 for later physician review, and, as needed, revision.

The Cross-Cultural Advisor (CCA) module is a key feature. Technically it is based on the NLP engine's capability for recognizing and flagging clinical content that requires special attention beyond what the NLP system can independently provide. In this case, the flags are associated with warnings related to subject matter that is known to have either cultural sensitivities for patients in the target language group or that is difficult to translate into the target language. Options that the CCA could present to the physician for any particular flag would include warnings with explanation of the sensitivity, pre-formulated queries or informational presentations that are designed to mitigate any misunderstandings, or advise that a human interpreter be involved. In cases where the services of a human interpreter are called for, the CCA identified topic can be used to select, when available, an interpreter with training or skills appropriate to the case at hand. This can be particularly useful when Video Medical Interpretation (VMI) capabilities are used and there is a pool of remote interpreters from which to select.

Triggering of the CCA is not simply a matter of recognizing key words in the discourse. This would lead to many spurious invocations of the CCA. As shown in Figure 1, the CCA is triggered only from elements of the physician's discourse. Although other trigger profiles can be defined, current triggers relate to specific patient conditions and circumstances, and to specific treatment modalities or some combination of the two. For example, a mention that the patient is pregnant would not in and of itself trigger the CCA. If, however, the patient is pregnant, is from a culture that attaches different significance and family roles and responsibilities with regard to child bearing, and the patient is experiencing fear or family pressure, or if the physician is anticipating the use of a procedure, say an epidural, that is not familiar in the patient's native culture, then the CCA would be triggered.

2 Demonstration Objectives

The initial and primary focus of our work has been mapping clinical speech to CDA2. The application of the system to medical interpretation grew out of this work due to the availability of SNOMED-CT (and other clinical nomenclatures, e.g. the International Classification of Diseases) in multiple languages. The CCA was a direct outgrowth from the flagging facility that A-Life uses in its medical coding applications. Direct translation, i.e. that which is not via the CDA2/SNOMED-CT mapping, is accomplished using third-party software. Given this background and the current state of development, the demonstration will focus on the following objectives. Demonstration of:

Clinical speech to CDA2/SNOMED-CT.

Transforming CDA2/SNOMED-CT to query form.

Translating CDA2/SNOMED-CT.

Triggering the Cross-Cultural Advisor.

For a more thorough written treatment with examples and references, the reader is directed to the full paper. [Heinze, Turchin, Jagannathan: 2006]

References

Heinze DT, Turchin A, Jagannathan V. 2006. Automated Interpretation of Clinical Encounters with Cultural Cues and Electronic Health Record Generation. *Proceedings of the HLT/NAACL 2006 Workshop on Medical Speech Translation.* Brooklyn, New York.

A Multi-lingual Decision Support Prototype for the Medical Domain

David Dinh
osTechnology Pty. Ltd.
Health Technology Solutions
david.dinh@ostechnology.com.au

Dennis Chan
PST Research Group
Voice Solutions Developer
dennis.chan@pstresearch.info

Jack Chen
PST Research Group
Health Solutions Developer
jack.chen@pstresearch.info

Abstract

In this paper, we are proposing a multilingual prototype that can effectively collect, record and document medical data in a domain specific environment. The aim of this project is to develop an electronic support system that can be used to assist asthma management in an emergency department.

1 Introduction

Speech technology has the ability to generate resource and time savings within a hospital environment. Recording and managing patient data from non-English backgrounds can be achieved successfully through the implementation of a multilingual voice system and a standardised electronic medical decision support system such as ACAFE (ACAFE 2006) described in Section 5.3. By implementing the ACAFE standardized protocols together with a voice system, we are able to assist in the first stage of the clinical pathway in the treatment and management of Asthma (see illustration of Stage 1 in figure 3).

In this demonstration description, we are proposing a multi-lingual voice system based on a standardized patient management system called ACAFE that can effectively collect patient data in electronic format. The combination of the two systems would make it easier to assist in the recording and documentation of vast amounts of information whilst overcoming communication and efficiency barriers. This data can then be aggregated and analyzed after the event to assist with clinical and performance measures. This makes effective use of emergency department resources while providing the emergency staff with immediate access to important patient information.

2 Objectives

To show how quality health care can be delivered in a complex multilingual hospital environment with the aid of an electronic decision support system such as ACAFE.

3 Demo Description

Our demo prototype integrates a voice recognition system together with the ACAFE system described in more detail in section 5.3. Our voice recognition prototype relies on data extracted from the standardized treatment protocols that have been based on research by ACAFE (ACAFE et al., 2006). These standardized protocols form the basis of our system-patient interaction to the medical subdomain (Starlander et al., 2005).

Since our system is heavily driven by ACAFE, we have been able to minimize the requirement for an open range of questions that require translation. As a result, we only require the use of the grammar-based language model (GLM) that has been implemented using Nuance's speech recognizer (Nuance 2005), and not a statistical language model (SLM).

The standardized protocols require no manipulation or changes in tense as the ACAFE system is essentially a decision support tool. The flexibility of the decision support tool allows the clinician to make the final decision and vary any responses or inputs. Hence the range of questions our multilingual system poses to the patient is also standardized and limited. With the smaller set of questions it is feasible for translation to occur via direct ACAFE to 'target-language' mappings (subject language to many variations of a target-language).

The use of GLMs over SLMs for medical speech translation has been proven to provide higher translation accuracy (Rayner et al., 2004, Rayner et al., 2005). We expect that by combining the higher accuracy levels of recognition through the use of GLMs with a limited set of possible questions for a particular medical sub-domain, we can achieve an improved translation success rate.

Currently, our system requires the Overseer (such as a nurse) to specify the patient's native language (in our example Chinese Mandarin) and problem sub-domain (in our example asthma). From there, the Overseer can either speak a question as defined in the protocols contained within the ACAFE system (using English), or select one using the terminal. The question is then rendered using recorded audio (TTS is used as a fall back strategy) and played to the patient. Once the patient responds verbally or physically (e.g. nod of the head), the Overseer is required to enter that response into the system.

The Overseer is capable of viewing reports that detail a particular patient's responses prior to further analysis/treatment, or they can view statistical reports. As a proof of concept, the Overseer can generate a statistical report that details patient background precipitating factors (numbers of respiratory tract infections, cold weather, exercise and dust/pollens)

4 Suggested Scenario

The triage nurse will identify the patient's native language to enable the correct voice system translator. The voice system will translate the standardized asthma management plan questions into the patient's native language.

Patient will answer each question in their native tongue. The voice system will convert this information into the ACAFE system format. When each question has been answered, the ACAFE system will store the answers and the voice system will then follow through to the next ACAFE question.

Upon completion of the set of ACAFE based questions the voice system will then provide a review of the questions with answers in the ACAFE system in either English or the native language. A voice recording will also be stored to play back for future reference.

Triage refers to the answers that have been collated in the ACAFE system via the assistance of the voice system. This information can be understood by all emergency team staff as the voice system has translated the answers of the patient into English according to the standardized management answers.

The Emergency Department now has a precompiled list of patient information compliant with Stage 1 of the clinical pathway contained in the ACAFE system to help assist in the treatment of asthma, without having to worry about communication difficulties between patient and medical staff.

4.1 Demo script

Triage Nurse: "Hello, what pains or difficulties are you experiencing?"
Patient: "Understand English no good, asthma…"
Triage Nurse: "Can you confirm your language, Mandarin or Cantonese?"
Patient: "Chinese, mandarin."
Triage Nurse: "OK, what I will do now is use a special machine to ask a few simple questions, you can just answer yes or no, it will ask the questions in mandarin so you can understand better. OK, here we go… "
Triage nurse then activates the voice system which goes through the set of ACAFE based questions in mandarin.

Figure 1: High-level view of user ACAFE interaction

5 System Architecture

5.1 Overview

Figure 2 illustrates a component view of the design for our prototype system. The Overseer acts as an overriding authority for the ACAFE Decision Support component, providing interpretations of the Patient's native language, medical problem sub-domain, and as a failover, the Patient's responses (both verbal and physical) to the questions asked.

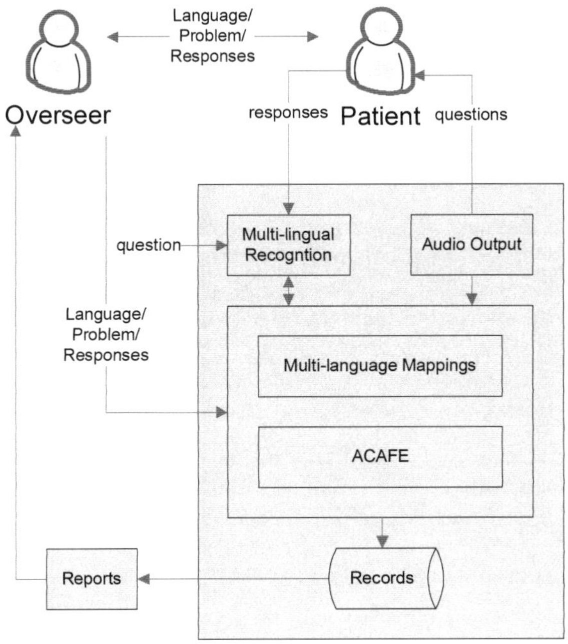

Figure 2: Component overview of the System Architecture

5.2 System Components

The following section outlines each component shown in the Overview diagram (Figure 2).

Audio Output – Renders questions (as required by the Decision Support) in the Patient's native language using recorded speech, or Text-to-Speech (TTS) if the recorded speech is not available.

Multi-lingual Recognition – The majority of questions posed to the Patient are in the form of yes/no questions. As such, the recognition of the Patient's utterance needs only to recognize basic responses in the Patient's selected native language.

ACAFE – Provided with the medical sub-domain (e.g. asthma/breathing difficulties), specifies questions according to a standard set of diagnosis questions.

Records – Records Patient responses to Questions (both textual and audio representations), final outcome, and statistics that are used for both individual Patient reporting and statistical reporting.

Reports – Provides individual Patient reporting (i.e. native language, medical sub-domain, responses to questions, and final outcome) and statistical reporting for the use of measuring the relationship between asthma and the precipitating factors.

5.3 Asthma Decision Support

ACAFE is an electronic interface for the Emergency Department that provides clinicians with a decision support tool to assist in the management and treatment of asthma. The system incorporates clinical decision support based on current evidence and guidelines that is simple to access, adaptable to the needs of the clinicians working in the ER and is capable of being integrated with existing medical databases.

The system's core focus lies in clinical pathways for the treatment of asthma. This is shown in Figure 3 below. A clinical pathway in the medical sense is a decision tree based on clinical assessment that guides the management and further investigation of a patient with a particular clinical problem. This decision tree has been based on consensus guidelines and institutional protocols based on the best available evidence for the management of asthma.

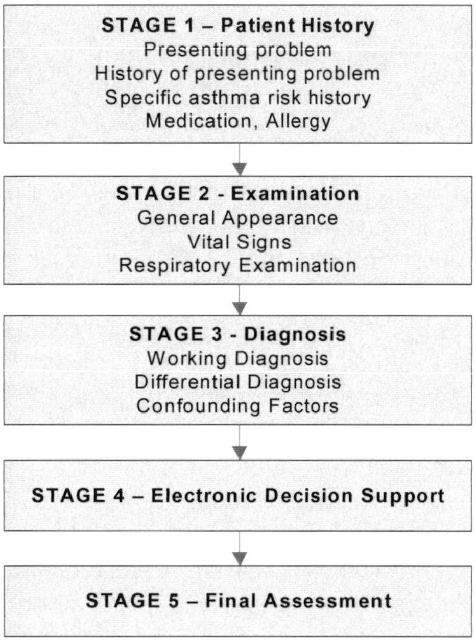

Figure 3: The ACAFE clinical pathway

In the ACAFE system the clinical pathway is represented by the information required to ascertain the severity of asthma to decide on a list of further investigations, consultations and medication orders. The clinical pathway outlines the means through which the system can advise the doctor on the optimal asthma management care plan.

At this stage, our voice system will be integrated with stage 1 of ACAFE's clinical pathway, in particular the history/information collection side of things.

6 Conclusion

We have shown that the ACAFE system with the assistance of our voice system can capture the information required to assist clinicians better manage the treatment of asthma in an emergency department. In capturing this data, the ACAFE and voice system incorporates the clinical pathways and decision support in the workflow of the doctor. In this demonstrator paper, we proposed a system that:

Relies on ACAFE by providing an electronic standardized protocol for the treatment of asthma.

Allows multi-lingual support thereby increasing communication between medical staff and patients during information collection and follow-up review after the patient has been discharged.

Increases efficiency by automating how information is collected by assisting in the recording and documentation of vast amounts of information while also streamlining the update of data electronically into the patient medical system.

Acknowledgements

We would like to thank osTechnology for their input relating to the ACAFE system development. We would also like to thank the emergency department team at Canterbury Hospital for their assistance and expert advice in the field of asthma patient care and information collection.

References

ACAFE Research Project and Development team, http://www.ostechnology.com.au/acafe/our_team.html. As of January 2006

Nuance, http://www.nuance.com. As of 8 December 2005.

M. Rayner, P. Bouillon, N. Chatzichrisafis, B. A. Hockey, M. Santaholma, M. Starlander, H. Isahara, K. Kanzaki, Y. Nakao (2005). *A Methodology for Comparing Grammar-Based and Robust Approaches to Speech Understanding*, In *Proceedings of Eurospeech-Interspeech, 4-8, September, 2005, Lisboa, Portugal.*

M. Rayner, P. Buillon, B. A. Hockey, N. Chatzichrisafis, M. Starlander (2004). *Comparing Rule-Based and Statistical Approaches to Speech Understanding in a Limited Domain Speech Translation System.* In *Proceedings of TMI 2004, Baltimore, MD UA, 2004.*

M. Starlander, P. Bouillon, N. Chatzichrisafis, M. Rayner, B.A. Hockey, H. Isahara, K. Kanzaki, Y. Nakao, M. Santaholma (2005). Breaking the Language Barrier: Machine Assisted Diagnosis using the Medical Speech Translator. In *Proceedings of the XIX Internation Congress of the European Federation for Medical Informatics MIE, 28 August - 1 September, 2005, Geneva, Switzerland.*

IBM MASTOR SYSTEM: Multilingual Automatic Speech-to-speech Translator *

Yuqing Gao, Liang Gu, Bowen Zhou, Ruhi Sarikaya, Mohamed Afify, Hong-Kwang Kuo, Wei-zhong Zhu, Yonggang Deng, Charles Prosser, Wei Zhang and *Laurent Besacier*
IBM T. J. Watson Research Center, Yorktown Heights, NY 10598

ABSTRACT

In this paper, we describe the IBM MASTOR, a speech-to-speech translation system that can translate spontaneous free-form speech in real-time on both laptop and hand-held PDAs. Challenges include speech recognition and machine translation in adverse environments, lack of training data and linguistic resources for under-studied languages, and the need to rapidly develop capabilities for new languages. Another challenge is designing algorithms and building models in a scalable manner to perform well even on memory and CPU deficient hand-held computers. We describe our approaches, experience, and success in building working free-form S2S systems that can handle two language pairs (including a low-resource language).

1. INTRODUCTION

Automatic speech-to-speech (S2S) translation breaks down communication barriers between people who do not share a common language and hence enable instant oral cross-lingual communication for many critical applications such as emergency medical care. The development of an accurate, efficient and robust S2S translation system poses a lot of challenges. This is especially true for colloquial speech and resource deficient languages.

The IBM MASTOR speech-to-speech translation system has been developed for the DARPA CAST and Transtac programs whose mission is to develop technologies that enable rapid deployment of real-time S2S translation of low-resource languages on portable devices. It originated from the IBM MARS S2S system handling the air travel reservation domain described in [1], which was later significantly improved in all components, including ASR, MT and TTS, and later evolved into the MASTOR multilingual S2S system that covers much broader domains such as medical treatment and force protection [2,3]. More recently, we have further broadened our experience and efforts to very rapidly develop systems for under-studied languages, such as regional dialects of Arabic. The intent of this program is to provide language support to military, medical and humanitarian personnel during operations in foreign territories, by deciphering possibly critical language communications with a two-way real-time speech-to-speech translation system designed for specific tasks such as medical triage and force protection.

The initial data collection effort for the project has shown that the domain of force protection and medical triage is, though limited, rather broad. In fact, the definition of domain coverage is tough when the speech from responding foreign language speakers are concerned, as their responses are less constrained and may include out-of-domain words and concepts. Moreover, flexible casual or colloquial speaking style inevitably appears in the human-to-human conversational communications. Therefore, the project is a great challenge that calls for major research efforts.

Among all the challenges for speech recognition and translation for under-studied languages, there are two main issues: 1) Lack of appropriate amount of speech data that represent the domain of interest and the oral language spoken by the target speakers, resulting in difficulties in accurate estimation of statistical models for speech recognition and translation. 2) Lack of linguistic knowledge realization in spelling standards, transcriptions, lexicons and dictionaries, or annotated corpora. Therefore, various different approaches have to be explored.

Another critical challenge is to embed complicated algorithms and programs into small devices for mobile users. A hand-held computing device may have a CPU of 256MHz and 64MB memory; to fit the programs, as well as the models and data files into this memory and operate the system in real-time are tremendous challenges [4].

In this paper, we will describe the overall framework of the MASTOR system and our approaches for each major component, i.e., speech recognition and translation. Various statistical approaches [5,6,7,8] are explored and used to solve different technical challenges. We will show how we addressed the challenges that arise when building automatic speech recognition (ASR) and machine translation (MT) for colloquial Arabic on both the laptop and handheld PDA platforms.

2. SYSTEM OVERVIEW

The general framework of our speech translation system is illustrated in Figure 1. The general framework of our MASTOR system has components of ASR, MT and TTS. The cascaded approach allows us to deploy the power of the existing advanced speech and language processing techniques, while concentrating on the unique problems in speech-to-speech translation. Figure 2 illustrates the MASTOR GUI (Graphic User Interface) on laptop and PDA, respectively.

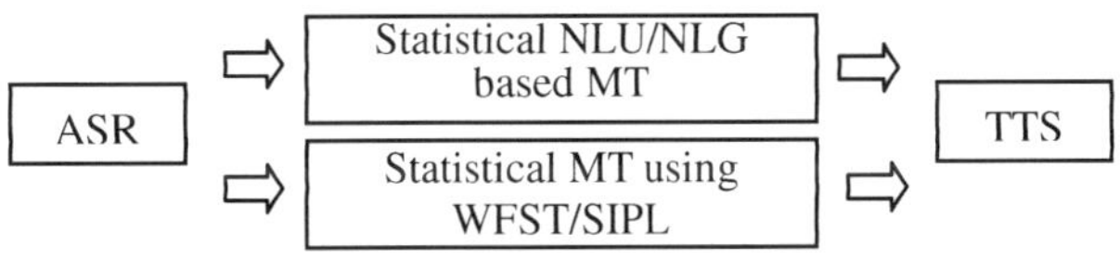

Figure 1 IBM MASTOR Speech-to-Speech Translation System

Acoustic models for English and Mandarin baseline are developed for large-vocabulary continuous speech and trained on over 200 hours of speech collected from about 2000 speakers for each language. However, the Arabic dialect speech recognizer was only trained using about 50 hours of dialectal speech. The training data for Arabic consists of about 200K short utterances. Large efforts were invested in initial cleaning and normalization of the training data because of large number of irregular dialectal words and variations in spellings. We experimented with three approaches for pronunciation and acoustic modeling: i.e. grapheme, phonetic, and context-sensitive grapheme as will be described in

* Thanks to DARPA for funding

Figure 2 IBM MASTOR system in Windows XP and Windows CE

section 3.A. We found that using context-sensitive pronunciation rules reduces the WER of the grapheme based acoustic model by about 3% (from 36.7% to 35.8%). Based on these results, we decided to use context-sensitive grapheme models in our system.

The Arabic language model (LM) is an interpolated model consisting of a trigram LM, a class-based LM and a morphologically processed LM, all trained from a corpus of a few hundred thousand words. We also built a compact language model for the hand-held system, where singletons are eliminated and bigram and trigram counts are pruned with increased thresholds. The LM footprint size is 10MB.

There are two approaches for translation. The concept based approach uses natural language understanding (NLU) and natural language generation models trained from an annotated corpus. Another approach is the phrase-based finite state transducer which is trained using an un-annotated parallel corpus.

A trainable, phrase-splicing and variable substitution TTS system is adopted to synthesize speech from translated sentences, which has a special ability to generate speech of mixed languages seamlessly [9]. In addition, a small footprint TTS is developed for the handheld devices using embedded concatenative TTS technologies.[10]

Next, we will describe our approaches in automatic speech recognition and machine translation in greater detail.

3. AUTOMATIC SPEECH RECOGNITION

A. Acoustic Models

Acoustic models and the pronunciation dictionary greatly influence the ASR performance. In particular, creating an accurate pronunciation dictionary poses a major challenge when changing the language. Deriving pronunciations for resource rich languages like English or Mandarin is relatively straight forward using existing dictionaries or letter to sound models. In certain languages such as Arabic and Hebrew, the written form does not typically contain short vowels which a native speaker can infer from context. Deriving automatic phonetic transcription for speech corpora is thus difficult. This problem is even more apparent when considering colloquial Arabic, mainly due to the large number of irregular dialectal words.

One approach to overcome the absence of short vowels is to use grapheme based acoustic models. This leads to straightforward construction of pronunciation lexicons and hence facilitates model training and decoding. However, the same grapheme may lead to different phonetic sounds depending on its context. This results in less accurate acoustic models. For this reason we experimented with two other different approaches. The first is a full phonetic approach which uses short vowels, and the second uses context-sensitive graphemes for the letter "A" (Alif) where two different phonemes are used for "A" depending on its position in the word.

Using phoneme based pronunciations would require vowelization of every word. To perform vowelization, we used a mix of dictionary search and a statistical approach. The word is first searched in an existing vowelized dictionary, and if not found it is passed to the statistical vowelizer [11]. Due to the difficulties in accurately vowelizing dialectal words, our experiments have not shown any improvements using phoneme based ASR compared to grapheme based.

Speech recognition for both the laptop and hand-held systems is based on the IBM ViaVoice engine. This highly robust and efficient framework uses rank based acoustic scores [12] which are derived from tree-clustered context dependent Gaussian models. These acoustic scores together with n-gram LM probabilities are incorporated into a stack based search algorithm to yield the most probable word sequence given the input speech.

The English acoustic models use an alphabet of 52 phones. Each phone is modeled with a 3-state left-to-right hidden Markov model (HMM). The system has approximately 3,500 context-dependent states modeled using 42K Gaussian distributions and trained using 40 dimensional features. The context-dependent states are generated using a decision-tree classifier. The colloquial Arabic acoustic models use about 30 phones that essentially correspond to graphemes in the Arabic alphabet. The colloquial Arabic HMM structure is the same as that of the English model. The Arabic acoustic models are also built using 40 dimensional features. The compact model for the PDA has about 2K leaves and 28K Gaussian distributions. The laptop version has over 3K leaves and 60K Gaussians. All acoustic models are trained using discriminative training [13].

B. Language Modeling

Language modeling (LM) of the probability of various word sequences is crucial for high-performance ASR of free-style open-ended coversational systems. Our approaches to build statistical tri-gram LMs fall into three categories: 1) obtaining additional training material automatically; 2) interpolating domain-specific LMs with other LMs; 3) improving distribution estimation robustness and accuracy with limited in-domain resources. Automatic data collection and expansion is the most straight-forward way to achieve efficient LM, especially when little in-domain data is available. For resource-rich languages such as English and Chinese, we retrieve additional data from the World Wide Web (WWW) to enhance our limited domain specific data, which shows significant improvement [6].

In Arabic, words can take prefixes and suffixes to generate new words which are semantically related to the root form of the word (stem). As a result, the vocabulary size in Arabic can become very large even for specific domains. To alleviate this problem, we built a language model on morphologically tokenized data by applying morphological analysis and hence splitting some of the words into prefix+stem+suffix, prefix+stem or stem+suffix forms. We refer the reader to [14] to learn more about the morphological tokenization algorithm. Morphological analysis reduced the vocabulary size by about 30% without sacrificing the coverage.

More specifically, in our MASTOR system, the English language model has two components that are linearly interpolated. The first one is built using in-domain data. The second component acts as a background model and is built using a very large generic text inventory that is domain independent. The language model counts are also pruned to control the size of this background model. The colloquial Arabic language model for our laptop system is composed of three components that are linearly interpolated. The first one is the basic word tri-gram model. The second one is a class based language model with 13 classes that covers names for English and Arabic, numbers, months, days, etc. The third one is the morphological language model described above.

4. SPEECH TRANSLATION

A. NLU/NLG-based Speech Translation

One of the translation algorithms we proposed and applied in MASTOR is the statistical translation method based on natural language understanding (NLU) and natural language generation (NLG). Statistical machine translation methods translate a sentence W in the source language into a sentence A in the target language by using a statistical model that estimates the probability of A given W, i.e. $p(A|W)$. Conventionally, $p(A|W)$ is optimized on a set of pairs of sentences that are translations of one another. To alleviate this data sparseness problem and, hence, enhance both the accuracy and robustness of estimating $p(A|W)$, we proposed a statistical concept-based machine translation paradigm that predicts A with not only W but also the underlying concepts embedded in W and/or A. As a result, the optimal sentence A is picked by first understanding the meaning of the source sentence W.

Let C denote the concepts in the source language and S denote the concepts in the target language, our proposed statistical concept-based algorithm should select a word sequence $\hat{A}$ as

$$\hat{A} = \arg\max_{A} p(A|W) = \arg\max_{A} \left\{ \sum_{S,C} p(A|S,C,W) p(S|C,W) p(C|W) \right\},$$

where the conditional probabilities $p(C|W)$, $p(S|C,W)$ and $p(A|S,C,W)$ are estimated by the Natural Language Understanding (NLU), Natural Concept Generation (NCG) and Natural Word Generation (NWG) procedures, respectively. The probability distributions are estimated and optimized upon a pre-annotated bilingual corpus. In our MASTOR system, $p(C|W)$ is estimated by a decision-tree based statistical semantic parser, and $p(S|C,W)$ and $p(A|S,C,W)$ are estimated by maximizing the conditional entropy as depicted in [2] and [7], respectively.

We are currently developing a new translation method that unifies statistical phrase-based translation models and the above NLU/NLG based approach. We will discuss this work in future publications.

B. Fast and Memory Efficient Machine Translation Using SIPL

Another translation method we proposed in MASTOR is based on the Weighted Finite-State Transducer (WFST). In particular, we developed a novel phrase-based translation framework using WFSTs that achieves both memory efficiency and fast speed, which is suitable for real time speech-to-speech translation on scalable computational platforms. In the proposed framework [15] which we refer to as Statistical Integrated Phrase Lattices (SIPLs), we statically construct a single optimized WFST encoding the entire translation model. In addition, we introduce a Viterbi decoder that can combine the translation model and language model FSTs with the input lattice efficiently, resulting in translation speeds of up to thousands of words per second on a PC and hundred words per second on a PDA device. This WFST-based approach is well-suited to devices with limited computation and memory. We achieve this efficiency by using methods that allow us to perform more composition and graph optimization offline (such as, the determinization of the phrase segmentation transducer **P**) than in previous work, and by utilizing a specialized decoder involving multilayer search.

During the offline training, we separate the entire translation lattice H into two pieces: the language model L and the translation model M:

$$M = Min\left(Min\left(Det\left(P\right)\circ T\right)\circ W\right)$$

where $\circ$ is the composition operator, Min denotes the minimization operation, and Det denotes the determinization operation; T is the phrase translation transducer, and W is the phrase-to-word transducer. Due to the determinizability of P, M can be computed offline using a moderate amount of memory.

The translation problem can be framed as finding the best path in the full search lattice given an input sentence/automaton I. To address the problem of efficiently computing $I \circ M \circ L$, we have developed a multilayer search algorithm.

Specifically, we have one layer for each of the input FSM's: I, L, and M. At each layer, the search process is performed via a state traversal procedure starting from the start state $\vec{s}_0$, and consuming an input word in each step in a left-to-right manner.

We represent each state **s** in the search space using the following 7-tuple: s_I, s_M, s_L, c_M, c_L, $\bar{h}$, s_{prev}, where s_I, s_M, and s_L record the current state in each input FSM; c_M and c_L record the accumulated cost in *L* and *M* in the best path up to this point; $\bar{h}$ records the target word sequence labeling the best path up to this point; and s_{prev} records the best previous state.

To reduce the search space, two active search states are merged whenever they have identical s_I, s_M, and s_L values; the remaining state components are inherited from the state with lower cost. In addition, two pruning methods, histogram pruning and threshold or beam pruning, are used to achieve the desired balance between translation accuracy and speed.

To provide the decoder for the PDA devices as well that lacks a floating-point processor, the search algorithm is implemented using fixed-point arithmetic.

5. CONCLUSION

We described the framework of the IBM MASTOR system, the various technologies used in building major components for languages with different levels of data resources. The technologies have shown successes in building real-time S2S systems on both laptop and small computation resource platforms for two language pairs, English-Mandarin Chinese, and English-Arabic dialect. In the latter case, we also developed approaches which lead to very rapid (in the matter of 3-4 months) development of systems using very limited language and domain resources. We are working on improving spontaneous speech recognition accuracy and more naturally integrating two translation approaches.

6. ACKNOWLEDGEMENT

The authors sincerely thank Drs. Yoshinori Tahara, Fu-hua Liu, Yongxing Li, Etienne Marcheret, Raimo Bakis, Ellen Eide, Burn Lewis, Tony Lee, Ossama Emam, and Lubos Ures for their help and contributions to the MASTOR S2S system.

7. REFERENCES

[1] Y. Gao et al, "*MARS*: A Statistical Semantic Parsing and Generation Based *Multilingual Automatic tRanslation System*," *Machine Translation*, vol. 17, pp.185-212, 2004.

[2] L. Gu et al, "Improving Statistical Natural Concept Generation in Interlingua-based Speech-to-Speech Translation," in *Proc. Eurospeech'2003*, pp.2769-2772.

[3] F.-H. Liu, "Robustness in Speech-to-Speech Translation," in *Proc. Eurospeech'2003*, pp.2797-2800.

[4] B. Zhou et al, "Two-way speech-to-speech translation on handheld devices," in Proc. *ICSLP'04*, South Korea, Oct, 2004.

[5] H. Erdogan et al, "Using Semantic Analysis to Improve Speech Recognition Performance," *Computer Speech and Language*, vol.19, pp.321-343, 2005.

[6] R. Sarikaya, et al, "Rapid Language Model Development Using External Resources for New Spoken Dialog Domains," in *Proc. ICASSP'05*, Philadelphia, PA, Mar, 2005.

[7] L. Gu et al, "Concept-based Speech-to-Speech Translation using Maximum Entropy Models for Statistical Natural Concept Generation," *IEEE Trans. Speech and Audio Processing*, vol.14, no.2, pp.377-392, March, 2006.

[8] B. Zhou et al, "Constrained phrase-based translation using weighted finite-state transducers," in *Proc. ICASSP'05*, Philadelphia, Mar, 2005.

[9] E. Eide et al, "Recent Improvements to the IBM Trainable Speech Synthesis System," in *Proc. ICASSP*, Hong Kong, China, 2003.

[10]Dan Chazan et al, "Reducing the Footprint of the IBM Trainable Speech Synthesis System," in *ICSLP-2002*, pp.2381-2384

[11]R. Sarikaya et al, "Maximum Entropy Based Vowelization of Arabic," Interspeech2006 (submitted for publication).

[12]L.R. Bahl, et al, "Robust methods for using context-dependent features and models in a continuous speech recognizer," in *Proc. ICASSP*, 1994

[13]D. Povey & P.C. Woodland, "Minimum Phone Error and I-Smoothing for Improved Discriminative Training," In *Proc. ICASSP*, Orlando, 2002.

[14]M. Afify et.al, "On the Use of Morphological Analysis for Dialectal Arabic Speech Recognition," Interspeech 2006 (submitted for publication).

[15]B. Zhou, S. Chen, and Y. Gao, "Fast Machine Translation Using Statistical Integrated Phrase Lattices," submitted to COLING/ACL'2006.

Association for Computational Linguistics
209 N. Eighth Street
Stroudsburg, Pennsylvania 18360